CLEAN WATER

Nature's Way to Stop Pollution

CLEAN WATER

Nature's Way to Stop Pollution

by
Leonard A. Stevens

A SUNRISE BOOK

E. P. DUTTON & CO., INC. | NEW YORK | 1974

Copyright © 1974 by Leonard A. Stevens
All rights reserved. Printed in the U.S.A.
First Edition

LIBRARY OF CONGRESS CATALOGING IN PUBLICATION DATA
Stevens, Leonard A.
Clean water; nature's way to stop pollution.
(A Sunrise book)
Bibliography: p.
1. Sewage irrigation. I. Title.
TD760.S84 1973 628'.36 73-79556
ISBN 0-87690-103-8

10 9 8 7 6 5 4 3 2 1

Published simultaneously in Canada by
Clarke, Irwin & Company Limited, Toronto and Vancouver

Outerbridge & Lazard, a subsidiary
of E. P. Dutton & Co., Inc.

Contents

v

Appendixes

Foreword

When I think of the land treatment process for sewage, which this book is about, I am reminded of a somewhat offhand remark made several years ago by David Dominick who was then Commissioner of the Federal Water Pollution Control Administration in the United States Department of the Interior. Commissioner Dominick, after listening to a presentation on the Muskegon, Michigan, proposal for an aerated lagoon, spray irrigation system for sewage treatment (discussed in Chapter 20), observed that his research people had told him breakthroughs were no longer possible in sewage treatment processes, but he thought Muskegon might prove them wrong.

The more I have thought about that remark, Muskegon being in my Congressional District, and having devoted a substantial amount of time to the question of sewage treatment systems in the subsequent three years, the more I am convinced that Commissioner Dominick's research advisers were wrong. It is not that the research people had misled the Commissioner, but rather that his advisers had neglected to inform him of the history and technological background available on the land treatment alternative. They were, themselves, unaware of the significance of this information. As Leonard Stevens so well documents in the following pages, land treatment systems are as old as civilized man yet as modern as the present intense and growing awareness of the need for protection of natural resources. But those systems have been virtually ignored by the sanitary engineering profession in its myopic "technologic fix" traditions. The Muskegon Plan and other projects discussed in this book are finally making a break with these unworkable traditions.

It is also interesting to recall the manifestations of strong, common sense support for this approach to sewage management on the part of the Nixon Administration. Upon being briefed by his staff on the Muskegon Plan, President Nixon wrote Dr. John R. Sheaffer, the father of the system, as follows: "The innovative system of water-borne waste disposal which you have developed for Muskegon County, Michigan, came to my attention recently. I understand that your imagination and dedication have led to the development of a new and promising approach to sewage disposal, and I want to commend you for your pioneering work in this vital field."

With such an expression of administrative support Muskegon was encouraged to develop its 10,000 acre, 43 million gallon-a-day land treatment system which began operations in May of 1973.

I am personally convinced that this method of sewage management will be recognized, through books such as we have here, not only as a means to clean our rivers and lakes of pollutants so that they may once again become national treasures, but also as a means to recover precious resources now wasted and to add to the total value of the product of human effort both in our own nation and abroad. More than one half billion dollars' worth of nutrients (now labeled pollutants) are lost each year to our rivers and lakes, when they really belong on the land. They belong where instinct and common sense both tell us they belong, on the farmer's field, enriching the soil and stimulating the growth of bountiful crops.

It is profoundly amazing to me that we have been able to ignore our instincts and common sense as long as we have. It is only our astonishingly good luck that we have been able to do so without bringing starvation and pestilence upon our heads. Had we poured our wastes into our drinking water in other times we could not have been so fortunate. Technology and modern medicine have saved us. But they cannot save us from the economic and esthetic costs of our foolishness. We cannot drink water from our lakes and streams without paying for expensive treatment processes. We cannot enjoy those bodies of water, as we might if they were not bacteria- and algae-ridden, foolishly used as part of our sewage treatment processes in violation of nature.

The essential logic of land treatment is its partnership with nature. Pollutants in water are fertilizers on land, if intelligently applied to the land.

Land treatment systems offer man the opportunity to benefit from nature, rather than to subject himself to a never ending need to "fix" the undesirable side effects of challenging that potent force. While land treatment systems are not new, our times demand a new perspective in understanding their significance, in understanding their potential, in understanding our need for the benefits they can produce.

James Russell Lowell told us years ago that: "New occasions teach new ideas; we cannot make their creed our jailor. They must forever onward sweep, and upward, who would keep abreast of truth. Nor attempt a future's portal with a past's outdated key."

The pages that follow open "a future's portal" with all of the excitement inherent in the discovery of nature's own key. Leonard Stevens challenges us to understand, and use, that key.

Guy Vander Jagt,
Member of Congress

PART 1

Converting a Blight to a Blessing

1

The Vital Miracle
of Soil and Plants

While the twin difficulties of water pollution and water shortages are discussed endlessly, few people ever deal with the main problem: *sewage,* the culmination of human wastes from wherever we live and work. Talk of sewage is abhorred and avoided. Except in scatological humor, the use of the word is indelicate, so in polite company sewage is called "waste water." Understandably, municipal officials concluded long ago that the only acceptable public policy for dealing with sewage was one of out-of-sight-and-out-of-mind. This was generally implemented by dumping sewage, raw or partially purified, into the nearest stream, with the hope it would disappear forever. But here sewage truly became a public plague by widely polluting our natural waters and greatly reducing our fresh supplies.

In recent years, however, the public outlook on sewage has slowly started to change as an increasing minority of citizens have sought answers to our worsening water pollution problems. Many have found that sewage need not always be the undiscussed scourge of towns and cities. Instead, they have been surprised to discover that the nation's waste waters contain resources, particularly valuable nutrients that are essential to green growth. These resources can be widely used to turn sewage from a blight to a blessing.

To realize this blessing we need a major change from our conventional approach to sewage treatment and disposal, which, under the penalty of water pollution, has generally worked outside of nature's system of treating and recycling the wastes from all living organisms, including man. By working closely with this natural system, we can actually put the resources of sewage back to work to our benefit. Instead of stimulating an aquatic growth that now causes much of our water pollution, such resources can be used to cultivate valuable plant growth on land, from grass to trees. In this way sewage can actually improve our environment while even producing economically valuable crops. Especially, it can help maintain the kind of attractive, beneficial open space that we so badly need to restore and preserve around urban America. And at the same time, we can truly solve our water crisis because we will be taking advantage of the oldest, most effective method for cleaning water, nature's method.

The need for such an approach to the nation's sewage was emphasized as the U.S. Congress overwhelmingly approved the Federal Water Pollution Control

Act Amendments of 1972, which states that ". . . it is the national goal that the discharge of pollutants into the navigable waters [which include all natural streams and lakes] be eliminated by 1985." The Act, a tough federal law, intended to clean up the nation's waters, presses for the reclaiming and recycling of waste waters with methods that take advantage of nature's way of cleaning water and reusing the ingredients that we now mislabel as pollutants.

The Act, which will provide billions and billions of dollars to clean up our polluted waters, calls for intensive citizens' action to see that the Congressional mandate is carried out across the nation. If the legislation and its money are to be used effectively, thousands of concerned citizens must turn their attention to the subject of sewage. There they face a choice between conventional waste water treatment methods based largely on technology and older but less conventional methods, now called "land treatment," based on nature's way of dealing with our wastes.

For most of this century scientists and sanitary engineers committed to technology have attempted to perfect purification processes that could remove the contaminants of raw sewage so a treatment plant's outflow (effluent) would be clean enough for disposal without polluting natural groundwaters (underground supplies) or surface waters. The application of these technological efforts is found in many communities with municipal sewage treatment plants.

Some are relatively simple "primary" plants that do little more than remove suspended solid materials from waste water, while more tenacious pollutants remain in the effluent. A better conventional system adds a "secondary" stage of sewage treatment that removes a larger percentage of the waste ingredients, particularly organic matter. But these expensive primary-secondary plants, which are difficult to maintain and operate, do not complete the sewage cleaning job, even when running at full efficiency. At best, "secondary effluent" still holds the threat of water pollution.

To meet that threat, today's sanitation experts, who are still heavily committed to a technological approach, are trying to develop a "tertiary" (third) stage of purification (also called "advanced waste water treatment") to finish the cleaning job that secondary effluent still requires. After a great deal of effort at tremendous cost, sanitarians are still pretty much at the pilot-plant level, and municipal tertiary treatment systems, of which there are only a few, have yet to prove themselves. In any event, whether such systems are primary, secondary, or tertiary, disposal of the treated contaminants has to be faced in the end and that's always a problem.

Meanwhile, the engineer's ultimate goal in purification was attained millenniums ago by the intricate natural system in common soil supporting live plants. Here lies a treatment process whose capabilities have yet to be fully recognized and understood. As one author has written, "A teaspoon of living earth contains five million bacteria, twenty million fungi, one million protozoa, and two hundred thousand algae. No living human can predict what vital miracles may be locked in this dot of life, this stupendous reservoir of genetic materials that have evolved continuously since the dawn of earth for about two billion years."

One "vital miracle" is nature's immense capability for cleaning and preserving great volumes of fresh water on and below the earth's surface. Consider how it worked in the floor of a primeval forest to maintain the pristine purity of woodland streams and ground-water reservoirs. The forest litter continually received tons and tons of waste from what was once described as "nature's city peopled with trees, birds, insects, shrubs, mammals, herbs, snails, ferns, spiders, fungi, mosses, mites, bacteria, and a myriad of other living forms." Their wastes, their droppings, their fallen leaves, twigs, and trunks, their dead, decaying bodies and parts were mixed with rain to become the sewage of "nature's city" in great volumes. If this sewage could have been analyzed by the modern sanitation engineer, he would have found most of the same ingredients that are now tagged as "pollutants" in human sewage. In the forest they weren't recognized as such, but were treated as resources, essential nutrients and organic materials to be returned to the growth processes from which they couldn't be spared. So as nature's sewage trickled down across the wooded slopes to a stream, or infiltrated the earth to the ground-water, the resources were recaptured and returned to work by the soil and plants, while the water was purified on the way.

The process was actually a link in one of nature's basic cycles that follows the essential nutrients of life from soil to plants and living organisms back to the soil via their wastes. Orientals called the cycle the "wheel of life," and today's school children learn about it as the "nutrient cycle." Under any name the administration of the cycle is bound to one of nature's most stringent rules, which can be stated in the single-word admonition: *Return!*

When man first arrived in "nature's city" his wastes became part of this great cyclical scheme, and the nutrients, which had come to him from eating plants and animals, were returned to the earth to continue their role in the wheel of life. When he began to discard his wastes more formally, man by necessity and by an innate sense of nature abided by the law of return. He restored his wastes to the land and advised his fellowmen to do the same. Evidence is found in the Bible (Deuteronomy 23:12–13): "Thou shalt have a place also without the camp, wither thou shalt go forth abroad; and thou shalt have a paddle upon thy weapon and it shall be, when thou wilt ease thyself abroad, thou shalt dig therewith and shalt turn back and cover that which cometh from thee."

In the eighteenth and nineteenth centuries, as expanding cities and towns with ample water supplies to carry off their wastes concentrated immense volumes of sewage, its disposal in rivers greatly increased the threat of water pollution and led to long, intensive studies by eminent commissions. They frequently concluded, as did one in 1865, that "the right way to dispose of town sewage is to apply it continuously to land." The advice was widely accepted, and the sewage from many cities and towns was applied to the land on municipal farms. When they were carefully laid out and well managed, raw sewage flowed into the acreage, pure water flowed out (on or below the surface), and a wide variety of crops were raised in between on the sewage-irrigated soil. As we shall learn, the farms were not only successful sewage treatment systems but esthetically attractive places where the workers often enjoyed better health and longer life than their friends in town, where the sewage originated. Subsequent research proved that nature's treatment systems are also highly effective for removing disease germs from waste water.

As time went on, the remarkable examples of "land treatment," which honored nature's law of return, were slowly engulfed by urban spread and seldom replaced. Only a few remain in the world today. The technological promise for waste water purification was followed instead, because at the time it seemed to offer neater, less expensive, more viable answers to the problem.

Some land treatment remained in the United States where arid conditions and/or economic pressures demanded use and reuse of water resources. But in recent decades raw sewage was seldom returned to the land because its application to the soil without creating nuisances or health hazards required the supervision of skilled managers who were unavailable in America. Instead, partially cleaned effluent from conventional treatment plants and other systems, which we will discuss later, was used to irrigate limited varieties of crops that were not eaten raw by humans. The effluent, which could be treated with chlorine to eliminate disease-carrying organisms, was of course valuable as life-giving irrigation water, but also it retained a good measure of the fertilizing resources of sewage. All told, however, the use of these resources was minor compared to the available amounts.

In recent times, when widespread water pollution has made it evident that conventional sewage treatment methods are insufficient, some authorities have gone back to the idea of land treatment as a viable alternative to technology, which is expensive to undertake and so far less capable than common earth and living plants. Even the National Academy of Sciences, in a 1966 publication on waste management control, recognized that the land "has a great capacity for receiving and decomposing wastes and pollutants of many kinds." And in 1971 the Subcommittee on Air and Water Pollution of the U.S. Senate was told by the eminent authority, Professor Daniel A. Okun, head of the Department of Environmental Sciences and Engineering of the University of North Carolina: "Much of the need for advanced waste treatment may be eliminated by applying wastewaters to the soil where this is feasible. Where such application may be useful in productive irrigation, the nutrient content not only is not a problem, but it constitutes a resource, having been estimated in California to be worth more than 5 cents per thousand gallons. Even where irrigation for the growth of crops is not contemplated, soils may be used for assimilating wastewaters."

The human benefits possible from land treatment of sewage are borne out in many ways, as the ensuing chapters will reveal. These benefits are evident in the pages of history which bear many instances of our more agrarian-minded ancestors respectfully recognizing the importance of abiding by natural laws for the well-being of both nature and man.

The rewards of land treatment are revealed by a

number of modern examples, where little-publicized efforts of certain municipalities and industries have worked to return waste waters to the soil—thus, in one stroke, controlling pollution, conserving water, and offering a great many people benefits derived from the economic and esthetic improvement of land and water for farms, golf courses, highway beautification, parks, wildlife sanctuaries, recreation lakes, fire belts, public gardens, and on and on.

The potential human values in land treatment are substantiated by modern research and pioneer projects which reveal that great volumes of sewage can be purified without danger to the public health and with much good to the public weal. But before sewage can be turned from a blight on our natural waters to a blessing on our land, increasing numbers of citizens must imitate nature in recognizing and understanding that our so-called wastes are really resources.

2

The Beneficial Ingredients
of Sewage

The valuable resources of sewage are twofold:

First, in this nation of serious water shortages, sewage and the pollution it causes keep vast quantities of water out of circulation for undue lengths of time. We now tie up some 250 billion gallons per day (over 100 gallons per person) and the consumption is expected to more than double by the end of the century.

Secondly, sewage contains uncounted tons of organic soil conditioners and fertilizers that can stimulate beneficial growth of many kinds. Today we dump it, with serious environmental consequences.

The classic advertisement for Ivory Soap might be applied to sewage water, which is more than 99 percent pure. The water is a carriage for waste materials that are only a fraction of a percent of the total flow.

As fresh water shortages have increased, the great volumes of water thrown away as sewage have been a source of some concern. For instance, back in 1954 three University of California engineers wrote: "To Water Works engineers and officials, and to enlightened citizens alike, the idea of rejecting used water simply because it has acquired the unsavory name of 'sewage' is becoming patently absurd. In what other field of human activity, they ask, is so valuable a vehicle employed to transport so small a burden; and in what other circumstance is the transporting vehicle discarded at the end of a single journey?"

Such statements often come from people urging that we tap the tremendous water resources of sewage to reduce heavy demands on precious natural supplies. Despite the potential, however, reclamation has been limited to a very small part of America's daily discharge of some 25 billion gallons of sewage. And then the reclamation is largely confined to the reuse of treated sewage water for irrigation in arid and semi-arid agricultural areas, and to industries needing large volumes of cooling water. Even so, the irrigation and cooling potentials in both instances have barely been touched, as we shall learn.

But the possibilities do not end with agriculture or industry. In a few revealing cases, the 99-percent-pure water in sewage has been reclaimed and reused by many people in surprisingly direct ways. For example, the drinking water for thousands of homes in one of America's most populous counties comes in part from sewage effluent filtered down through the earth—and thereby purified—to supplement the natural supplies for municipal wells. In one of the nation's largest, most beautiful city parks, people have been boating for many years on a lake supplied by treated sewage effluent. And not long ago the citizens of a small West Coast community started boating, fishing, and swimming in several new lakes where every drop of the water had just been reclaimed from the municipal sewerage system.

The fertilizers and soil conditioners of sewage which cling tenaciously to that fraction of a percent have been characterized by generations of sanitation authorities as culprits, pollutants difficult to remove and be rid of without causing a public nuisance. But this portrayal in itself has been part of the problem. It has kept many of us from realizing that some of the so-called contaminants of waste water are really valuable resources which could be used to our benefit rather than suffered as a curse.

Let's consider these wasted resources in a way that a knowledgeable gardener might think about them.

Among the so-called contaminants of sewage he would immediately recognize some elements essential to gardening, especially nitrates, phosphates, and potash. These compounds, the first two in particular, are widely discussed by sanitary engineers because they are among the most difficult, indeed, nearly impossible ingredients to remove in conventional sewage purification systems, and they are blamed for a lot of today's water pollution.

But the compounds are also a holy trinity that good gardeners worship as essential to plant growth. They are listed on fertilizer bags where the proportion of each is indicated by three numbers, like 5-10-5 (5 percent nitrogen, 10 percent phosphorus, 5 percent potash). Many gardening books stress that the triumvirate are the most important elements in soil, and the ones used up first. Among other things, they enhance the growth of leaves and stems, stimulate root development, help form seeds, and develop disease resistance.

The same fertilizers found in sewage can improve green growth as well if not better than comparable commercial fertilizers — indeed, the waste-water variety are organic and would probably be preferred over mineral fertilizers by today's increasing numbers of organic gardeners. In one experiment, sewage effluent from a town's primary-secondary treatment system was found equivalent to a 7-14-12 fertilizer (7 percent nitrogen, 14 percent phosphorus, 12 percent potash). When only two inches of such sewage effluent was applied to an acre of land per day for a month, the equivalent of a ton of 7-14-12 fertilizer was simultaneously delivered to the soil. A ton of comparable commercial fertilizer would have cost about seventy dollars at the time (the mid-1960's).

The sewage of a city or town discharges a tremendous tonnage of these valuable plant nutrients, considering the fact that in a year the average person contributes the following to sewage: nitrogen 7.5 to 8.5 pounds, phosphorus 2.0 to 2.5 pounds, and potash 4.0 to 6.0 pounds. This means that fewer than two dozen persons supply enough nitrogen, phosphorus, and potash annually to meet the fertilizer requirements for growing an acre of high-quality mixed crops. Furthermore, this potential resource is available without energy expenditures (for manufacture and transport) that may cause air pollution and deplete natural resources, as is the case with commercial fertilizer production.

When sewage is discharged with fertilizers into lakes and watercourses, they are eventually washed out to sea and lost to the land where they really belong. But in the meantime, when the volume of waste water is great enough, the fertilizing compounds stimulate the uncontrolled growth of algae and aquatic plants which can literally choke fresh water to death. The fatal process is known as "eutrophication." The fertilizers consigned to this unintended role thus become tagged as pollutants. The well-publicized demise of Lake Erie is a leading example of how this form of aquatic death has devastated America's natural waters.

The gardener who continues to review the contaminants of sewage will also find frequent reference to organic substances. The sanitation engineer works hard to remove them from waste water because they too are identified as extremely serious water pollutants. The organic matter is food for aerobic bacteria (the kind that need oxygen, as opposed to anerobic bacteria which do not). If too much such food is discharged with raw or treated sewage into fresh water, the resulting bacterial growth can, again, literally choke the water to death. It happens when the thriving biological life depletes water of the dissolved oxygen that helps maintain other organisms and plants essential to keeping the water fresh. This oxygen-reducing effect of the organic matter in sewage is referred to by engineers as "biological oxygen demand" (BOD), which is translated into a numerical measurement to indicate how much a body of water is polluted by such organic substances.*

The wise gardener, however, is always seeking organic matter which builds humus, one of the most important constituents of soil. Humus improves earth in a number of ways, enhancing the growth of plants. For example, it helps the soil absorb and use plant food, and resist erosion. But humus, even in the best-managed garden, disappears and has to be replaced continually. It can be added by using manure, peat moss, compost, and other materials, some of which can be purchased in bags. Where available, dried sewage sludge, an organic residue of solid materials settled out of waste water in treatment plants, is used by gardeners as a humus builder. Sewage water, which is not so available to the public, also contains organic matter that can build humus in soil.

Finally, the more sophisticated gardener studying waste water could find a number of constituents that are not so commonly discussed by sanitarians but are essential to green growth.

For instance, sewage has a certain amount of that

* BOD is expressed by a number determined in a common test. A water sample is analyzed for dissolved oxygen at the beginning and end of a five-day interval. The difference between the two results, expressed in parts of oxygen per million parts of water, reveals how much dissolved oxygen is being demanded by the bacteria, whose numbers depend in turn on the amount of organic matter they have to feed upon in the water. The test is thus an indication of how much organic material is polluting the water. A nice clear stream may have a BOD as low as 5 parts per million (ppm). The typical home sewage has a BOD around 200 ppm. The highly polluted waste water of a food processing plant may test as high as 1,000 ppm or more.

most common of garden products, lime (calcium). In waste water it's found in the form of carbonate. The good gardener knows that lime has to be continually replenished in high-organic soils, for it is taken up by plants and living microorganisms, and is washed away by rain.

The knowledgeable gardener would also be interested to learn that sewage contains a number of "trace elements," which are minute quantities of soluble material including iron, sulfur, sodium, boron, magnesium, iodine, manganese, copper, and zinc. Typical domestic sewage contains so little of these elements that they are not problems to municipal treatment plants, nor are they polluters of fresh water. Some industrial sewage, however, may include one or more of the elements in large, toxic quantities capable of destroying the biological action essential to sewage treatment systems and to keeping natural fresh water fresh. Pollution control authorities have concluded that industries with such toxic wastes must be ordered not to dump them into municipal sewage or natural water supplies.

But in the world of plant growers, the same trace elements in reasonable amounts are welcomed ingredients important to growth. They are sometimes called "micro-nutrients," and without them plants suffer. Lacking tiny amounts of iron, for example, a plant will fail to develop its green color because the element is a key to the essential production of chlorophyll.

The manner of looking at all these ingredients of sewage—whether they're seen and treated as contaminants or resources—has a lot to do with whether they shall serve a destructive or constructive role in nature. Our out-of-sight-out-of-mind psychology regarding human waste has forced these resources into the destructive role of polluting lakes and watercourses, rather than stimulating beneficial growth on land. Thus we break nature's basic law of return.

The fertilizing elements come to sewage from the land via the food chain, leading from the soil through plants and animals to humans, at which point they are found in our excretia. If we obeyed nature's law, the nutrients of human waste would be returned to the land so they could continue their essential, circular journey in the "nutrient cycle" or "wheel of life." But we break the law in disregard of the natural cycle by dumping unacceptable tons of nutrients with sewage into rivers, streams, lakes, and oceans. The penalty has been stiff. We have been sentenced to water pollution on a national and international scale.

Besides having the penalty lifted, should we start abiding by the rule of return, we could also expect a reward: green carpets of growth in fields, forests, parks, and many other land areas. It's possible, if we intelligently return our sewage nutrients to the soil with living plants instead of to the water with aquatic growth.

The proposal is not new. It was heard by our ancestors, whose agrarian minds appreciated the needs and benefits for returning human waste to the land. They often recognized that sewage held resources that could "swell the wealth of the nation." In *Les Miserables,* written in the 1860's, Victor Hugo complained that by disposing of sewage in the Seine,

"Paris casts 25 millions of francs annually into the sea. . . ."

For several decades the sewage of many great cities was returned to the land with remarkable demonstrations of the beneficial possibilities, but then land treatment was largely abandoned as society was changed by technologically minded people who worshiped machines and believed they could do all things. With the vow of clean water neatly and safely sanitized from sewage, people were unsold on nature's systems and persuaded to buy expensive factorylike methods that never did fulfill their promise—as one may witness on a walk beside most of America's once clear and beautiful rivers.

PART 2

The Amazing Purification Farms

3

The Sewage Difficulty
(*circa* 1890)

The following might have been written in the 1960's
or '70's, but it was set in print many decades earlier:
"Owing to the rapid growth of urban population during the past few decades, and the consequent increase
of pollution of streams from which water supplies are
obtained, the subject of sewage purification and its
relation to the purity of streams has attracted general
attention. In view of the large number of people concerned and the benefits to be derived from the dissemination of information on this subject, there is
probably no topic . . . which is of greater importance
to the country as a whole."

The paragraph appeared in 1897 as the introduction to the first paper of a series published by the
U.S. Geological Survey. The author, George W.

23

Rafter, had made an extensive personal survey of sewage treatment in Europe and America. He was one of several investigators trying to help town and city officials find the best methods of sewage disposal. Their research convinced them that, where possible, the most effective method was to apply public waste waters to land. This had the backing of Europe's most eminent committees and commissions, and through their travels the American investigators proved for themselves that taking sewage to the soil was the one sure way of purifying municipal waste water and preventing the pollution of streams.

Water pollution had been serious in Europe for many years, and concern was increasing in the United States, especially in the East where urban growth was multiplying the volume of sewage. Actually the difficulties could be traced back to some admirable sanitation improvements largely associated with Sir Edwin Chadwick, a famous English reformer. Chadwick was an early and successful advocate of using the public water supply for sewage disposal systems. He said that water for domestic purposes supplied to homes through one set of underground pipes should also be used to dilute sewage and flush it out through another system of pipes for disposal.

Many cities already had extensive sewerage systems, but they were limited to the drainage of storm water from the streets. Their use for domestic wastes was strictly forbidden. Reformers like Chadwick successfully fought this restriction, and many large communities opened their collection systems to domestic sewage as well as storm water. They became known as "combined sewers."

The drive to construct sewerage works was also motivated by terrible epidemics. Over fifty-three thousand English citizens died of cholera in 1848 and 1849, only six years before Sir John Snow of London showed that the disease could be spread when wells were polluted by domestic sewage. Fewer deaths quickly proved the value of the new sewerage systems. There was even a general increase in the average longevity of the English people coinciding with the first decade of sewerage reforms.

But the solution of one difficulty brought another. It was well described in 1867 by an English civil engineer, Baldwin Latham, when he gave "A Lecture on the Sewage Difficulty," in which he said, ". . . the lives and health of our citizens have been purchased at the expense of the rivers. It must be admitted that the purity of water is of great importance, and we shall not, as a country, derive the full benefit to be obtained from sanitary measures until our rivers have been freed from the abominations that have been poured into them. This matter which at present pollutes our rivers, when properly utilized, instead of being a bane to society, would tend to swell the wealth of the nation. The early sanitary reformers of this country, knowing full well the baneful effects of retaining faecal and other decomposing matters in the midst of our populations, were led to look upon sewage as a nuisance to be got rid of as speedily as possible; so they poured it into the nearest river or watercourse, in the hope that it would be borne harmlessly to the ocean, and there be forever entombed from our sight. But nature rebels against the waste, and the matter is left to seethe on the banks of rivers, or is returned by the

tidal wave that rolls on our shores." Latham, an advocate of land disposal, went on to say that in throwing
the sewage back to the shore, the ocean was saying
"in the strongest terms" that man's waste water be-
longed on the land, not in the sea.

"The sewage difficulty" was one of the most investigated problems of the nineteenth century. All over
Europe official committees and commissions were
established to find ways of purifying the flood of sewage from the new collection systems. One of the most
extensive studies was made by the Sewage of Towns
Commission, which was appointed to find the most
profitable, beneficial means of using English sewage.
After a lot of study and travel the members settled on
land disposal and urged in several reports that wherever possible sewage water should be taken to the
soil for profits from agriculture and for the thorough
purification of the water. The English Society of Arts,
one of the nation's most respected scientific organiza-
tions, also concluded that the land held the answer to
the disposal problem.

The recommendations for land treatment from these
and other investigators were widely adopted, and in
the last quarter of the century sewage from many
European towns and cities was piped to farms for use
as combined irrigation and manure. By the end of the
century many American communities also began using
this system of sewage disposal.

Today it often surprises people to learn that, when
well managed, these farms of the late nineteenth and
early twentieth centuries were remarkably successful.
With soil, crops, air, and sunshine, they turned raw

sewage into a remarkably clean, clear grade of water that did not pollute natural groundwater supplies, lakes, or watercourses. Indeed, they accomplished what engineers subsequently failed to duplicate with technological systems. Moreover, the farms yielded profitable crops often grown in attractive "green belts" near the cities they served. Because of the vigorous growth, strangers riding trains in Europe could often tell when they were approaching a major city for the passing countryside suddenly appeared greener. As we shall learn, some of the farms are still successfully operated, but most were the victims of disinterest and bad management fostered by the false assumption that technology held the easiest way out of the sewage problems suffered by towns and cities.

The success of the early European ventures brought investigators like George Rafter from America, as United States urban growth caused the same "sewage difficulty" experienced in Europe twenty or thirty years earlier. They often returned home as staunch advocates of land treatment, and their reports became part of a growing debate as to how American cities should save their great rivers from the rising volumes of sewage.

Their opposition came chiefly from sanitation engineers recommending chemical plants which supposedly cleaned sewage well enough to keep it from polluting natural waters. The chemical systems were essentially a series of large tanks where sewage was dosed with a number of chemicals that turned the organic matter into "insoluble precipitates" which would settle and form a thick sludge at the tanks' bottoms.

The plants had machines for grinding and mixing the chemicals, for introducing them into sewage, and for removing the sludge so it could be carted away.

The proponents of land treatment argued that the chemical plants were insufficient and their adoption would only delay the day when America's rivers, streams, and lakes would be fouled by the partially treated effluent.

4

The Farms of Berlin, Paris, and London

Some of history's most effective sewage treatment systems were well-managed farms. They were agriculturally productive, attractive to visit, and healthy places in which to work and live. For half a century or so, beginning around the 1860's and '70's, they provided a high degree of sewage purification and water pollution control that the technology of this century has failed to attain. The most renowned farms served London, Paris, and Berlin.

The Berlin farms were models that attracted sanitarians from all over the world. They were particularly interesting to American officials because the German soil, climate, and other natural features were similar to those around communities of the northeastern United States, where the rising volume of sewage was

29

a serious threat to the rivers. A comprehensive report on Berlin that circulated in America came from an English engineer, Herman Alfred Roechling, who had visited the German metropolis in 1892 and observed the famous farms for some time.

When the sewage drainage system for Berlin was first proposed in 1861, city officials intended to remove only some of the solid materials and dump the remaining waste water into the river Spree which ran through the city, but the idea was rejected. Later the city council settled on sewage irrigation, and eventually Berlin's waste water was piped to city-owned farms north and south of the metropolis. By the time of Roechling's visit there were four farms on the sandy plains of northern Germany, totaling nineteen thousand acres, treating the sewage for the city's 1,578,794 people. On each farm sewage was pumped to a tall stand-pipe (reservoir) on the highest point of land, where they became landmarks visible for miles across the German countryside. The sewage, which was drawn from the reservoirs as required, flowed by gravity through pipes to the various fields on which two irrigation methods were in use.

In one, which became known as "overland flow," the waste water was released from ditches along the tops of gently sloping fields of grass. It would trickle slowly through plants down across the land to collection ditches at the bottom. At this point the effluent was remarkably clear and pure, as it was drained off the farms to nearby streams.

In the second method, large, flat terraces planted with row crops were enclosed by small earthen dikes.

Sewage, led to the terraces by pipes and open channels, was used to flood irrigate the crops. The water then slowly filtered down through the earth three or four feet to a series of buried pipes designed to collect and drain the purified effluent off to ditches that carried it to nearby streams.

Both types of irrigation had to be carefully managed to insure the intermittent use of each plot. A limit of a two- or three-thousand-gallon dose of sewage was maintained per day per acre to make certain that the soil had the benefit of some resting time every twenty-four hours. Gangs of workmen had the responsibility for applying sewage to the various fields at specified times night and day. Each of 134 "sewage men" was assigned to about fifty-nine acres.

"Everything in connection with the irrigation is done in military order," Roechling reported; "the day and night sewage-men parade at 6 A.M. and at 6 P.M. at an appointed place, generally in the yard of the principle homestead on the farm, whence, after calling of the roll, and the examination of kit and tools, they march each to his particular district. Every man carries in a tin case over his shoulder a book containing minute directions of what he is to do, and what punishment he will get if he does not carry out his instructions. He must fill up a form attached to the book stating what sluice-valves he has opened during his shift (all sluices being numbered), the time of opening and closing each, how many revolutions he gave to the spindle of each valve (an indication as to how much sewage passed through it), and what plots he irrigated. Every inspecting official has to enter his name on this form, giving also the time of his inspection, and the state in which he found things. At the end of the month these forms are sent to headquarters. To Englishmen it might seem that such arrangements were over-elaborate, but in Berlin they work remarkably well. The Author was present at one of the parades of the sewage-men, and he was struck with their smart military bearing, and when walking over the farms he was often impressed with their thorough comprehension of the work."

Roechling included these details to emphasize that strict management practices were the key to sanitary, esthetically acceptable, commercially successful results. Around 1882 careless management had left the Berlin farms open to serious public complaints which led to the militarylike management program. "It is this army of thoroughly trained sewage men," Roechling said, "that enables the authorities to cope most successfully with sewage-irrigation upon a scale that is at present without parallel."

The English engineer found over 80 percent of the farms under cultivation. They produced winter wheat, barley, and oats, as well as root crops and vegetables such as turnips, carrots, potatoes, and cabbages. In most cases the harvests were heavier on the sewaged land than comparable unsewaged acreage. The farms also produced exceedingly abundant crops of grass averaging over twenty-three tons of hay per acre. The ordinary, unirrigated meadow of the area averaged only 2¼ tons per acre. The miles of roads running through the Berlin farms were lined with seventy-one thousand fruit trees which added to the produce. And

one farm operated a profitable "home for horses
. . . requiring rest and a change of air."

The German sewage engineers irrigated arable
land practically year round. In the cold months, the
incoming sewage, which was appreciably warmer than
the atmosphere, usually kept the soil free enough of
frost for irrigation to proceed. When hard frosts prevented soil infiltration for brief periods, the sewage
was temporarily stored in shallow, man-made ponds of
five to twenty-two acres. In warmer months the same
basins would catch and temporarily store large, sudden volumes of storm water so it could be used later at
a reasonable rate.

When the capital costs and year-round operational
expenses were totaled, the Berlin sewage farms could
not compete favorably with conventional, privately
owned farms. However, the German sewage officials
weren't out for profits. They looked upon their acreage
not primarily as farms but as highly effective sewage
treatment systems.

They were proud of their excellent, rich soil, for
example, because analyses revealed that it was doing
a remarkable job at removing phosphorus and potash.
The sewage required annually to irrigate an acre arrived with 586 pounds of phosphorus and 884 pounds
of potash but the effluent left with only 8 and 56
pounds respectively. (In terms of removal from the
sewage this represented 99 percent phosphorus and
94 percent potash.) Soil tests revealed a general build-
up of these elements, and German scientists were
analyzing plants to see how much of the two substances was being used in the growth process.

Roechling reported that the effluent was constantly monitored to check the farms' effectiveness as purification systems. For a decade Dr. E. Salkowski, a professor at the Royal University of Berlin, had made some three hundred effluent tests on the four farms in all seasons. Many of his testing methods are still applied in the latter half of the nineteenth century for the analyses of organic matter and nitrogen, phosphorus and potash. Also bacteria counts were made to determine the effluent's health hazards. As a whole, the extensive testing revealed an "excellent" quality of effluent that would certainly not pollute groundwater supplies or surface watercourses.

From his own observations the Englishman reported: "On his several visits to the farms, the Author took particular notice of the condition of the subsoil effluent and of the effluent streams as apparent to the senses. The water was in almost every case perfectly clear and transparent, and only in one or two instances could he notice a slight earthy smell. In Gross Beeren [one of the farms] he was informed that, during 1886, eighty good-sized pike were caught in the effluent ditches. Had he not known that the streams contained the effluent from the farms, the Author would have been unable to detect it, as there is absolutely no trace of contamination."

The purified water, Roechling said, "cannot be distinguished by the senses from clear spring water. All those who visited the farms bear testimony to the absence of any smell in the fields, and only in one or two places near a sluice-outlet could any unpleasant smell be perceived when the sluices were opened."

The best proof of the farms' remarkable sanitary condition was in a four-year tabulation of all cases of sickness and deaths among the average number of 1,580 people living on the four farms. This figure included some 968 men of whom 850 were described by Roechling as " 'misdemeanants,' men impoverished in health by bad and irregular habits of life."

"The [human] material being such," the author commented, "it would not have been surprising if a heavy death-roll were found to exist on the farms; but . . . just the reverse is the case." He continued: "Nearly every report mentions that in no recorded case of death was it possible to trace that it had any connection with sewage-farming, and the report for 1888–89 contains this remarkable passage: 'Special stress must be laid upon the fact that not one case of typhoid fever (*typhus abdominalis*) has been recorded during the year upon any of the farms, though the eastern and northern portions of the city were visited at the commencement of the year with a severe outbreak of this fever; this being the first epidemic of any extent since the establishment of the farms.' "

These records were more remarkable in light of a revelation made by one of the farms' head gardeners. Sewage workers, he said, were forbidden to drink the farms' effluent, but the rule was constantly broken. The workmen liked to eat their lunches out in the fields where they never hesitated to dip up effluent for a drink of water to wash down the food.

While Berlin's farms were the leading models of how effective the soil and crops could be for sewage purification, similar evidence came from all over Europe that it would be hard to match this kind of system. The reports were often written by engineers, like Roechling, who were comparing land treatment with the new chemical treatment systems.

Paris was frequently visited by the investigators. Around 1870 the French metropolis began piping its sewage to the plain of Gennevilliers, north of the city, where it was given to farmers for irrigation. At first a public outcry arose over the practice of spreading the sewage and about the produce sold from the farms. But the objections subsided, and the popularity of sewage irrigation increased. By 1880 the Paris waste water was so much in demand that the city began selling it to farmers, who produced a wide variety of crops including garden vegetables which were sold in Paris.

George W. Rafter, who wrote the paper on sewage irrigation for the U.S. Geological Survey, visited the farms in 1894, and wrote: "The crops grown under sewage irrigation at Gennevilliers have been successful in the highest degree. They comprise absinth, artichokes, asparagus, beans, beets, cabbages, carrots, celery, kohl-rabi, cucumbers, leeks, melons, onions, parsnips, peppermint, potatoes, pumpkins, spinach, tomatoes, turnips, clover, rye grass, mangolds, wheat, oats, and Indian corn. The market-garden produce yields abundant crops. Indian corn has also an exceedingly good growth here, the stalks attaining a height of from nine to ten feet."

About fifteen acres of land were purchased on the plain of Gennevilliers by the city of Paris for its "Model Garden," which was used many years by the sanitation department to develop the best methods of sewage irrigation. For example, the weights of seeds and their yield were precisely recorded in relation to the amount of sewage they received. Also, careful rainfall records were maintained and compared with the amounts of sewage application to learn how the farms could be best managed under various weather conditions. The information was passed on to farm proprietors to help insure good crop yields and a high degree of sewage purification.

Many studies of effluent from the plain of Gennevilliers consistently confirmed its remarkably high quality. To assure the readers of his report that the Paris sewage was purified, Rafter wrote as follows: "The Author visited the Plain of Gennevilliers on a rather warm day early in December 1894. The effluent was bright and sparkling, and as he was exceedingly thirsty after a long walk he had no hesitation in dipping up the water of the effluent channel and drinking it, the gentleman accompanying him having made the positive assurance that no harm would result therefrom. Thus far there is no reason to believe that the effluent from sewage irrigation from the Plain of Gennevilliers may not be used as drinking water regularly with impunity."

Rafter also collected statistics on the health of the French sewage farmers, and as was true in Berlin and elsewhere, they didn't suffer from their trade. The American wrote: "So extensively has the evidence on this point multiplied from the different places where sewage utilization works have been carried out that we may conclude here . . . that properly conducted sewage utilization is not in any degree prejudicial to the health of either the people engaged in it or those living in the vicinity."

Bacteria counts made by the Mountsouris laboratory in Paris further proved that the Gennevilliers farms truly purified the water. While raw sewage in the Paris mains had 29,454,000 bacteria per cubic centimeter, counts in the effluent from the farms were as low as 5,380 per cubic centimeter.

Some seven thousand acres of the Paris farms are still used today for treating raw waste water from the French metropolis. They remain as cooperative ventures between the city and private farmers. Crops include root celery, green beans, carrots, turnips, leeks, corn, artichokes, parsley, and cabbage, which are sold in the city for human consumption. The land is also used to grow nursery stock, shrubs and trees, for sale in the city. The managers of the project (some having been there over forty years) cannot recall any substantiated case of the farms causing a health problem. In only one episode in a long history were they blamed for such a problem, but the charges were found untrue. The farms had complete public acceptance, and the personnel believed there was no better way of treating sewage, even in the 1970's.

Europe's other notable sewage farms of the last century included several that served the Borough of Croydon, a suburb of London. One of the farms, with 425 acres under irrigation, was laid out on a series of gentle, uniform slopes, and the sewage water trickled down through the crops to collection ditches, where it left the farm as sparkling clear water. Purification worked all winter long. If the ground was covered with ice, the sewage flowed underneath it. At this time, however, the liquid had to cross a greater expanse of land to attain the same degree of purification possible in summer months.

When George Rafter visited this English farm, he was surprised to see fine country homes along the highways in and around the irrigation fields. "In the course of a walk about the farm," he stated, "a stop was made at the residence of a gentleman whose garden plat of three or four acres is surrounded on three sides by the irrigation fields of this farm. As to the question of nuisance, the owner stated that no nuisance had ever been created by the sewage irrigation that he had considered in any way objectionable, although there were occasionally slight smells in warm weather. When the place was first occupied by him, over twenty years ago, he had rented, but on the expiration of the lease he had bought the freehold of the property, because it suited him, and he had not considered in making such purchase that the value of the property was in any degree injured by the presence of the sewage farm. He also said that a neighboring cow yard, where a considerable number of cows were kept for the purpose . . . of feeding sewage-grown produce, was regularly a source of much more serious effluvian nuisance to the neighborhood than ever came from the farm itself. At the time of the author's visit, in October, absolutely no smell of any sort could be detected in any part of the irrigated fields."

In England in the late 1800's the Royal Agricultural Society held a competition with prizes for the

best sewage treatment farms. In choosing the winners
the society intensively studied the farms around London, and the research added more evidence that land
treatment was superior to the competing method of
chemical treatment — characterized by one commen-
tator as "the milk and water plan of putting in a small
quantity of sulphate of iron and permanganate of
Potash."

The competition judges were most surprised by
their health findings, which were described by Baldwin
Latham, the eminent sanitary engineer, as follows:
"One of the things they had to inquire into was the
health of the people living upon the sewage-farms, and
the statistics brought out the remarkable fact that, of
those living upon the farms, a large number of whom
were children, . . . the rate of mortality was under 4
per 1,000, showing that in the act of utilizing the waste
products, which otherwise were so baneful to life, the
process, properly carried out, became of advantage to
the public health."

The American investigators sailed back to the
United States with all these glowing reports in favor
of land treatment, and they were used in a lot of lively
debates as to whether nature or man-made technology
would become the key to sewage treatment.

5

U.S. Land Treatment, Formal and Fortuitous

The first attempt at land treatment in America is said to have been at the Augusta, Maine, State Insane Asylum about 1872. Some seven thousand gallons per day of the hospital's waste water were mixed in a large tank with straw, leaves, muck, and other absorbent materials. The overflow was then led off to irrigate a field where the asylum's patients raised three crops of "fine hay" per season and cultivated a thriving vegetable garden. Many other asylums adopted the practice, and for good or bad, land treatment was often associated with such institutions.

The first municipal farm in the United States was laid out in 1880 and 1881 near the model town of Pullman, Illinois, about fourteen miles south of Chicago. The land was underlain with common agricultural drain tiles at a depth of about four and a half feet, to carry away the purified sewage effluent after it had filtered down through the black prairie alluvian. The farm was patterned after the best European models, and for many years its crops returned an 8 to 10 percent profit for the venture. The most successful were vegetables, like cabbage, cauliflower, and celery, which were sold in Chicago.

However, the farm eventually failed at its main purpose of sewage purification, and observers like George Rafter blamed poor management for unnecessarily overtaxing the soil with too much waste water. Eventually the urban spread of Chicago engulfed the farm, and Pullman adopted other methods of sewage treatment.

Many American towns and cities used land treatment with varying success, but few could compare to the great European farms. In 1899 Rafter wrote a second paper for the U.S. Geological Survey where he said: ". . . thus far American farms, generally speaking, have not realized their full agricultural capacity by reason of defects in management. We need to develop here, as there have been abroad, a class of sewage-farm managers who, having made a specialty of this method of farming, shall be not only informed as to details but qualified to meet emergencies.

"As regards sewage-purification works owned and operated by towns, it may be said that the entire separation of the management of the works from local politics will be the first necessary step toward increased efficiency."

However, a few successful farms in America proved

late last century that land treatment could work here as well as anywhere. One example was Brockton, Massachusetts, where thirty acres of land were divided into twenty-three well-designed beds of sand and subsoil for the filtration of sewage on a carefully managed schedule. In the winter months the beds acted strictly as filters to clean waste water as it seeped down through the soil to the natural groundwater. Rows of parallel furrows cut over the length of the beds prevented the soil from freezing and blocking the sewage infiltration. In the growing season a variety of vegetables — beans, tomatoes, sweet corn, turnips, and peas — were raised in the beds and sold in town. The farm's best customers included Brockton's leading hotels. The Massachusetts State Board of Health analyzed the farm's effluent from test wells and concluded that the city's sewage was being purified to drinking water quality.

But then at the turn of the century the advocates of land treatment were opposed by some persuasive engineers promoting the sale of chemical treatment systems. They were attractive, especially in our eastern states where sewage was a big problem, where land was in demand for urban development, and where society, thriving on its industrial prowess, was increasingly awed by the idea that there was nothing a machine couldn't do. People were inclined to buy the chemical plants, even though they were expensive and not always what they were claimed to be.

For instance, Worcester, Massachusetts, which had dumped raw sewage into the Blackstone River, installed a chemical purification plant after being forced to clean up its waste water by the Massachusetts legislature in response to complaints from Millbury, located downstream from Worcester. Despite the expensive plant, the pollution of the Blackstone continued, and Millbury sued Worcester. The long trial carried testimony from the nation's top authorities on sewage treatment and stream pollution, and on December 14, 1896, Worcester was again ordered to clean up its sewage. Chemical purification was clearly not enough, and the city was back in the market for an effective means of sewage purification.

But in these days the enemies of land treatment proved formidable. They could find plenty of ammunition at the mismanaged farms so prevalent in America. Even the noxious reputation of one badly managed farm with its mishandled sewage could be used by their opponents to cause voters to reject any proposed sewage irrigation project. Public fear was also easily kindled by blaming the farms for spreading diseases, even though there was plenty of evidence to the contrary.

With these adversities the farms were gradually abandoned as urban development overtook the existing ones. Also, as we shall find out, some new kinds of treatment technology came along in place of the chemical systems, and the quality of effluent seemed acceptable enough for dilution in America's large, vigorous rivers without fear of pollution.

By the 1930's land treatment with raw sewage applied to farms was all but gone. One of the last in the East was at Vineland, New Jersey. It finally fell to mismanagement and was abandoned prior to World War II.

However, a great deal of American farmland still received millions of gallons of sewage without the onus of being a sewage farm. They were simply irrigation farms, although the water was sewage, sometimes as raw as could be. Indeed, the practice continues to this day, and some United States farms enrich their fields with sewage receiving only minor treatment.

The irrigation water came from rivers where the flow was maintained at times as much, or more, by sewage discharge as by the natural fresh water supply. For that matter, during the driest seasons some of the river sources of irrigation water were literally open sewers. Irrigation companies or private farmers diverted the sewage-supported waters to irrigation ditches and on to their fields. In a number of cases, the diversion occurred right at a community's sewage outfall, the point of discharge for all of its waste water. From these outfalls in many instances raw sewage was guided directly to irrigation ditches before it was diluted at all with the river water.

In 1939 the U.S. Department of Agriculture published an extensive bulletin, "Sewage Irrigation as Practiced in the Western States," and considerable space was devoted to the use of natural water supplies carrying "raw sewage or effluents in varying stages of treatment or subsequent natural purification." The author remarked, "Whether the . . . use of such stream water for irrigation in a given case is or is not sewage irrigation may involve some fine distinctions."

The bulletin offered numerous examples of this casual sewage farming. It was happening along the Yakima River in Washington, which carried raw and partially treated sewage from several cities. The Boise and Snake Rivers of Idaho supplied a lot of raw sewage for irrigators. The sewage of Ogden, Utah, made up over half the downstream flow of the Ogden River which was being pumped to many farmers by two irrigation companies.

In over one hundred areas surveyed in the western states, sewage irrigation was cultivating a wide variety of crops, small grains, field corn, alfalfa, berries, and fruits. The report also told about sewage-irrigated vegetables, some of which were being raised in commercial quantities, although state laws were generally opposed to the practice. The plantings included tomatoes, beans, onions, peas, cabbage, beets, cucumbers, sweet corn, carrots, pumpkins, radishes, squash, lettuce, okra, peppers, spinach, eggplant, chard, celery, parsnips, rutabagas, turnips, asparagus, Jerusalem artichokes, parsley, and rhubarb.

The government survey found that crops irrigated with sewage waters generally had a higher yield and that the soil structure was improved by the continued application of raw sewage or partially treated effluent. The report was less certain about the dangers to natural groundwater supplies. There were fears that haphazard, careless irrigation with sewage endangered water supplies, but the report was certain that where effluent was carefully applied, it actually augmented underground water, and it could be used without fear of contamination.

By whatever name it assumes, this fortuitous kind of sewage irrigation is still practiced in the West, and it happens in small and large irrigation ventures. In a Texas town recently a farmer with little more than enough land to raise a bale or two of cotton suddenly began winning a coveted prize for harvesting the first bale of the season in his county. His surprised fellow farmers with a lot more experience and equipment wondered how he was beating them — until the secret got around town. The local primary-secondary treatment plant was located just a few yards upstream on a river from the prize-winner's farm. But the plant was out of order, so that raw sewage flowed through, untreated, into the river. Without a word to anyone the farmer rigged up a pump and irrigated his small cotton patch with the rich river water. His cotton came up fastest of all in the county, and he walked away with the annual prize for the first bale.

By the time the town was forced by the state to restore its treatment system, the officials had become aware of the values of their sewage effluent, and they arranged with another farmer with many acres to utilize all the community waste water for irrigation. Now he's first in line with the early cotton.

The value of a sewage-laden river in irrigation country quickly becomes evident in the courts if the waste water supply is cut off for some reason. This happened on the South Platte downstream from Denver in the 1960's, and the court cases were not settled until 1972.

For a long time, during certain seasons, the South Platte coming out of Denver carried a large percentage of sewage effluent with only the minor purification provided by primary treatment. At the edge of the city, several large irrigation companies owned a common headgate which diverted river water to hundreds of miles of irrigation ditches serving over a thousand farmers raising sugar beets, corn, barley, and hay. But then Denver formed its Metropolitan Sewage Disposal District encompassing thirteen area municipalities. The district collects all the primary-treated effluent around Denver and pipes it to a huge new secondary treatment plant on the South Platte — some distance downstream from the common headgate. The irrigation companies, which had enjoyed the river supply rich in sewage nutrients for many decades, not only found themselves low on water but learned that the remaining flow would now lack the city's effluent because it was bypassing the headgate.

The companies went to court in 1967 to claim riparian rights to the "normal" supply cut short by the new district. The irrigators wanted the effluent flow restored by having the court order the city to pump the secondary discharge of the new plant back upstream to the common headgate.

The desirability of the effluent was emphasized by a sidelight to the case. Ten other irrigation companies with headgates on the South Platte below the new plant entered the case as intervenors in behalf of the city. "Our loss had been their gain," said one of the irrigators who sued the city, "and they wanted to keep every bit of the extra water that was now coming their way from the new treatment plant."

Eventually the complainants lost their case, but arrangements were made with Denver to pump a share of effluent back upstream to their headgate — however, the companies had to pay the pumping costs. The farmers along the affected ditches still receive the rich effluent that both irrigates and fertilizes their fields, but now it costs more.

For many years western farmers have practiced sewage irrigation without design, illustrating that the soil can absorb large volumes of waste water, obviously with great benefit to the user and minimum harm to the public, even when the plan is carried out in a relatively haphazard fashion. They also made a point that emerges repeatedly in the history of sewage. When people are motivated, whatever the reasons — from financial gain to the esthetic improvement of the environment — they have few qualms about reusing their own wastes in safe, unobtrusive ways.

6

The Melbourne and Metropolitan Board of Works Farm (*circa* 1970's)

As interest in land treatment of waste water was revived in recent years, United States scientists and engineers, reviewing the history of sewage irrigation, were of course impressed by how effective the municipal farms of the past were at sewage purification. If such a farm remained in operation anywhere, they believed it could offer valuable data for the development of modern land treatment systems. For example, soil that had been irrigated with raw sewage for decades could tell a great deal about the possibilities for long-term land treatment, even though the modern proponents intended to irrigate with treated effluent instead of raw waste water. There turned out to be such a farm, although it was a long distance from home: Melbourne, Australia. In 1972 a delegation of

American engineers, scientists, and government officials flew to Australia to study the system, the Melbourne and Metropolitan Board of Works Farm.

A visitor arriving at the farm, about a half-hour drive from the city of 2.5 million people, finds it difficult to believe that here is the metropolitan sewage purification system treating some 100 million gallons of domestic and industrial waste water per day. Actually the visitor sees a magnificent cattle farm, indeed, one of the largest in a country noted for big cattle ranches. There are tree-lined country roads from which one can view broad, beautifully lush pastures totaling some forty-two square miles with herd after herd of handsome beef cattle cared for by stockmen, the Australian version of the American cowboy. The farm soil, once considered inferior agricultural land, is now acclaimed as the country's most productive grazing territory. Besides attracting sewage authorities from many countries, the Board of Works Farm has brought ranchers from Texas and other notable cattle lands to marvel at the stock, which provide Australians with their choicest beef. They include sleek white-faced Herefords, glossy black Aberdeen Angus, and Shorthorns, totaling some nineteen thousand animals. On a summer day, the visitor may also see some of the forty thousand sheep grazing in deep green pastures on what was once a barren plain.

To see the Board of Works Farm from within its boundaries makes one forget that its basic purpose is to purify the sewage of one of the world's great cities. It was established in 1893 on 8,847 acres of treeless plain along the shores of Port Phillip Bay southwest of Melbourne. The farm's main crop was grazing grass, which was irrigated with the sewage from the homes and industries of the city. As Melbourne and its waste water volume grew, the farm was expanded to fulfill its central purpose of sewage treatment. It now has 26,809 acres. "Today, after seventy years of successful operation," the Board of Works proclaims, "the Farm is one of the show places of Australia and a splendid example of practical economics of turning waste into wealth."

Two methods of land treatment are employed on the farm. One is used in the late spring, summer, and autumn, when the waste water is applied to over ten thousand acres of pasture divided into large plots and graded so the water will spread evenly over the soil and filter down into the earth. The sewage is applied in accordance with a carefully designed program. First, the cattle are removed from the pastures to be irrigated, and then the equivalent of four inches of waste water is released upon the land from channels that convey the liquid around the farm. As soon as the water percolates into the soil, and the surface dries out enough that the cattle's hoofs will not wreck the turf, the stock are returned to graze. In eighteen days or so the procedure is repeated.

About two-thirds of the water disappears into the area's natural groundwater or is lost to evaporation and transpiration. A third migrates laterally on the clay base below the topsoil to drainage ditches that lead it off for disposal in Port Phillip Bay. This effluent is clear, safe from a health standpoint, and so low in pollutants that it meets the stiff requirements demanded in Great Britain for effluent discharged into inland streams used for drinking water. Instead of fouling Port Phillip Bay, the pollutants, so-called, remain to improve the land — as a Board of Works' paper explains:

"The waste water as received contains valuable plant foods, both in suspension and in solution — particularly nitrogen, phosphoric acid, potash, and lime, as well as traces of many other elements. The amount applied annually to each acre of pasture contains nitrogen and phosphoric acid equivalent to 1,950 lbs. of sulphate of ammonia, 2,300 lbs. of blood manure and 670 lbs. of superphosphate, all of which, of course, are common commercial fertilizers. These fertilizers naturally promote a prolific growth of pasture, and heavy grazing is necessary to keep it in check. . . ."

The cattle, in addition to providing beef that helps pay for the farm, play an essential role in the purification process. They are described by the management as "grass mowers" that remove (eat) a share of the sewage nutrients taken up by the vegetation. When the beef is sold, the Board of Works literally turns its "waste into wealth," as much as $500,000 a year (in Australian dollars). The farm's sheep and their wool may add another $250,000 annually. These commercial enterprises reduce the city's sewage purification costs, which, incidentally, may vary from year to year according to how well the farm products do in the marketplace. One recent year, for example, the per capita cost of sewage treatment was 67 cents, but the next it was up to $1.07. Of course, the farm income depends somewhat on how well its animals meet the nation's health tests for beef cattle. The Board of Works Farm beef remains among the best in the country.

The second method of land treatment is called "grass filtration." The technique replaces pasture irrigation when it becomes impractical in the cold, wet winter months from May to August. The irrigators then turn to another part of the farm and some 3,472 acres divided into plots, each graded with a gentle slope leading down to a drainage ditch at one side. The plots are planted with Italian rye grass. The waste water, after standing briefly in settling ponds to remove the solid materials (sludge), is allowed to trickle through the thick grass down the slopes, and it is thus purified. Some forty-eight acres of land will treat a million gallons of sewage per day. It remains effective through most of the winter period, but the efficiency begins to wane as the waste water is transferred back to the pasturelands. However, the graded grasslands restore themselves as purifiers during the intervening months prior to being irrigated. In this period they receive a single "mowing" by the farm animals who, at the same time, stomp the natural grass seeds back into the ground to help with the next season's growth.

The farm also includes numerous ponds, called "lagoons," which are designed to catch and hold huge peak loads of waste water. With heavy rains, storm water entering the city sewerage system may more than double the average daily flow of waste water to the farm. When the inflow rises above a certain level, the system of some 520 miles of distribution channels automatically releases the waste water to the lagoons. The excess is held there for six weeks in which time nature, with air and sunshine, cleans the water to a fairly high degree of purity by a biological process that we shall explain in a future chapter. Actually, the lagoons represent a third sewage treatment technique on the farm.

Still, the land and grass filtration methods are most effective as sewage purifiers. This claim has been supported over the years by soil analyses which have also shown how the farmland has been improved. The following tabulation of resource elements, compared in parts per million to the farm soil, indicates how the earth has gained nutrients from sewage irrigation:

	Before Irrigation	After Irrigation 12 years	26 years
Nitrogen	1,260	2,620	5,000
Phosphorus	450	1,700	2,500
Potash	1,540	8,010	10,920
Lime	600	3,200	3,900

Some of the Board of Works Farm's most important assets can't be tabulated in figures. For example, it has been a good place to work for three generations of Australians. The farm once employed several hundred workers, many of whom lived within its borders. Because of mechanization, the number is now nearer one hundred. Of all the seven thousand people in various jobs with the Board of Works, the hundred at the sewage treatment farm are the healthiest employees of all, as proven by the board's records of sick leave.

The farm is also acclaimed for preserving valuable open space between Melbourne and Geelong to the southwest, a rural area that could easily have been engulfed by urban sprawl from both communities. Situated as close as it is to the large metropolitan area, the farm attracts its city neighbors who enjoy leisurely drives through the irrigated grazing lands on a number of public roads. After a short ride from the city, people can see the real equivalent of Australian back country, complete with cattle, sheep, and stockmen. Local visitors include lovers of mushrooms, which grow around the borders of the grazing lands among the hundreds of thousands of trees that have been planted over the years on the once treeless plain. Along with the trees and all the growth promoted by irrigation have come an abundance of animals and birds, so the Board of Works Farm was led into playing another role; it was designated as a wildlife sanctuary, to the joy of many nature lovers, especially ornithologists, who constantly visit the farm.

Such people are certain to be asked if the Board of Works Farm smells bad. The answer is no, certainly no worse than any farm. This observation was even backed up by an official Committee of Inquiry appointed by the state government. The committee made a long study of odors through many inspection trips to the farm and surrounding towns. The members concluded that there were odors causing complaints to be leveled against the farm, but investigations invariably proved that the smells came from sources outside the large sewage treatment area.

The farm manager, S. S. Searles, and resident engineer, C. F. Kirby, wrote of their huge ranch, saying that it provided four public benefits: sewage purification, Australia's best beef, a wildlife sanctuary, and open space.

PART 3

The Need for an Alternative:
A History

7

Dilution and/or Treatment Plants

Anyone trying to understand or explain why we need an alternative to our conventional, machinelike sewage treatment systems can use a brief history of America's water pollution by sewage. It might start with two events occurring less than two months apart in Chicago in 1892.

The first, which happened on September 3, was the ground-breaking for the "Seventh Engineering Wonder of the United States": the Chicago Sanitary and Ship Canal. While the canal would open up shipping between the Mississippi and Chicago, it would also allow diversion of the city's sewage from Lake Michigan to the mighty Mississippi and on down to the sea. The construction of the canal was a major victory for sanitarians, who claimed that much of America's sewage could be harmlessly diluted by our large and vigorous rivers. It seemed like the permanent answer to one of Chicago's most difficult problems.

The city had been plagued by sewage from the mid-nineteenth century, when a new collection system carried its waste water into Lake Michigan—also the source of Chicago's water supply. The lake had seemed big enough for both purposes, until August 2, 1885, when more than six inches of rain fell in a few hours. The water scoured the city streets, sewers, and catch basins, and it pushed a foul black mass of pollutants far out into Lake Michigan, beyond the city's drinking water intakes, which then brought disease and death back into the metropolis. The incident forced the formation of the Metropolitan Sanitary District of Greater Chicago and construction of the Sanitary and Ship Canal.

Some 8,500 workers took 8 years to dig the 23-mile channel, which was 160 feet wide and 24 feet deep. On January 16, 1900, control gates were finally opened to allow water from Lake Michigan to flow down the new canal, and it was soon diluting the city's sewage, which was led to the channel by huge sewer lines. Lake Michigan was relieved of Chicago's waste water, but the canal, and the river systems beyond it, were virtually turned into open sewers, to the horror of people downstream. In 1900 the State of Missouri took the State of Illinois and the Chicago Sanitary District to court to prevent the windy city's "undefecated and unpurified" sewage from being delivered to the Mississippi. The case, which lasted six years, was finally dismissed by the Supreme Court of the United States, and the dilution method, despite its obvious drawbacks, had a future in America.

The second event in 1892 was the introduction of a new kind of sewage treatment plant at the World's Columbian Exposition which opened in Chicago on Columbus Day. The plant was designed to purify the exposition's six to nine million gallons of sewage daily. After the waste water passed through a mechanical system of vertical tanks with filters, the effluent was disposed of in Lake Michigan. The man in charge was Allen Hazen from the Massachusetts State Board of Health Experiment Station on the Merrimack River at Lawrence, which was famous for thousands of experiments on sewage treatment. Hazen monitored the exposition plant, and reported that the purification was not very efficient. Back at Lawrence the search continued for treatment technology that could produce sewage effluent that wouldn't cause water pollution when dumped in a river.

Both events in Chicago were previews of the two main approaches that the nation's sanitarians would follow in trying to solve America's growing sewage problem in the twentieth century. Their thinking would be dominated by dilution and man-made treatment plants. Neither worked very well, as proven by today's widespread water pollution.

To begin with, sanitarians all over America emulated the Chicago decision to rely on dilution for sewage disposal. It was an attractive, relatively inexpensive way of putting a city's sewage out of sight and out of mind. But it meant that river after river was taken over as part of a sewage treatment system, and that's the way most of them remain today.

The great example is *"Dat Ol' Man River,"* who *"jes' keeps on rollin' "* with millions and millions of gallons of raw sewage. Of 90.5 million gallons dumped into the Mississippi at New Orleans, some 65 million gallons are discharged raw and the other 25 million receive only primary treatment which just removes solids. Memphis still honors the discredited dilution theory and so do the sanitary departments of numerous other communities with millions of people along the great Mississippi. Ol' Man River is in serious trouble. Lesser rivers succumbed long ago. That early victory for the dilution proponents devastated one river after another. Sooner or later there was always more sewage than this or that stream could dilute. Pollution was the result.

When a city's sewage was more than a river could take, the usual answer was a treatment plant that would hopefully purify the waste water enough to allow the dilution process to continue. This happened over and again across the nation. And, to go back to our leading example, it happened in Chicago.

By 1908, right after the city's Sanitary District started dumping raw sewage in the new drainage canal and polluting downstream waters, sanitation officials decided that sewage treatment was necessary, and they began seeking and developing the methods. A 1930 decision of the U.S. Supreme Court reduced the amount of lake water that could be diverted to the canal, dilution became less effective with less water, and treatment was now more needed than ever. Since then, Chicago has built many plants, but the search for a fully efficient treatment method still continues in the windy city and all across the nation. The effort is a prolonged story of scientists and engineers trying, but not succeeding, to duplicate what nature can do with soil and plants.

The chemical methods of sewage purification, which were so vigorously promoted in the late nineteenth century, often deceived people. The chemicals were effective in coagulating and settling solid materials enough to produce a remarkably clear effluent, but tests for biological oxygen demand (BOD) showed it was still high in pollutants. Furthermore, huge quantities of the settled solids (sludge) were a major disposal problem by themselves. By 1910 most of the chemical plants had been abandoned in favor of other methods of purification.

The chemical idea was revived briefly in the United States around 1930, but it suffered the same old problems, plus some new economic difficulties. And in recent years, chemical treatment has had a second revival, but it's too early to report on the outcome, although there are still problems which indicate that the rebirth may not lead to a long, happy life.

The most important and durable advances in sewage purification came as scientists discovered the role bacteria could play in breaking down the wastes in sewage. The process was called nitrification (the formation of nitrates from nitrogenous compounds), and it was seen as the key to a purification system. In the beginning, two papers by an English scientist, Robert Warington, reported on a two-part experiment that revealed a lot about the nitrification of sewage.

First he boiled a dilute solution of urine to kill the bacteria, and then he prevented airborne bacteria from reaching it by filtering the air through cotton batting. Under these conditions nitrification failed to occur. But then he added a small particle of fresh soil to the sterilized solution, and in a short time the nitrification process started converting the ammonia of the urine to nitrate. The process worked best in the dark and at 99 degrees Fahrenheit. Finally, the English scientist reported that nitrification did not occur when the soil particle had also been boiled before the test. Warington concluded that the purifying action of soil had three effects:

• It acted as a simple filter removing undissolved solid matter from the water as it percolated into the soil.

• Certain amounts of ammonia and organic matter were simply given off and reduced by the soil.

• The ammonia and organic matter were acted upon by the native bacteria of the soil.

While Warington's work helped scientists understand how soil cleaned waste water, his experiments were the basis for a new kind of sewage treatment plant. Another contributing experiment was performed in France by a biologist, Jean-Jacques-Théophile Schloesing. After filling a long glass tube with marbles and sand, he allowed raw sewage to trickle through it. At first nothing happened, but then the muddy liquid emerging from the bottom cleared up. He attributed the purification to the work of living organisms that were being cultivated on the surface of the marbles and sand, and he proved his point by adding a little chloroform to the sewage water which killed the organisms and ended the clarification process.

The most extensive studies of the bacterial action on sewage were conducted at the Lawrence (Massachusetts) Experiment Station late last century and early this century. In a twenty-two-month period, over four thousand chemical analyses and untold numbers of microscopic examinations were made on various combinations of soil and sand to determine their filtration qualities. The experiments indicated that when purification occurred, the nitrifying organisms attached themselves to the grains of sand and pebbles, forming a thin film on the surfaces. The investigations also showed that once the organisms were functioning, the sewage was best treated if it stopped and remained briefly in contact with the film-covered surfaces. And the experiments further indicated that while the sewage was being detained, the most purification occurred when the whole porous area was impregnated with as much oxygen as possible.

Allen Hazen, who had run the treatment plant at the Chicago Exposition, published a report on the Lawrence experiments proposing that the design of a biological treatment system rested mainly on two factors: time and oxygen. Hazen claimed if sewage had enough time in contact with a filter's organism-laden surfaces, and if enough oxygen were forced through the filter's millions of interstices, a high degree of purification would occur.

From all the experiments came a filter that is still

a key element in the secondary stage of thousands of today's sewage treatment plants, namely the "trickling filter." It is usually a large, fairly deep, circular bed of stones over which sewage water is spread, often as spray from a long radial arm that slowly revolves just above the bed. As the system first goes into operation, a layer of bacteria soon forms on the stones, and the organisms feed upon the passing sewage, thus helping to purify it. Meanwhile, air is mechanically forced through the bed to add the essential oxygen.

While the trickling filter was being developed, after the turn of the century, more than a dozen other investigators were studying the possibilities of reducing the obnoxious features of sewage by "aeration" (blowing air through it), and eventually they hit upon the idea for a new, more efficient method of sewage treatment, the "activated sludge process." As the aeration experiments began, little attention was paid to the solid materials, the sludge, settling from the air-filled waste water. But then some English workers at Manchester collected the material and reintroduced it into the churning, bubbling sewage water of their aeration tanks. Surprisingly, it speeded up the purification process. Sewage could be treated in a matter of hours compared to days required by the trickling filter.

By World War I the activated sludge experiments were being translated into plant-scale operations to treat thousands or millions of gallons per day in a continuous flow. Since then the theory and application have been refined, and today's activated sludge process is found in the secondary stages of the best primary-secondary treatment plants.* They are the systems most likely to be recommended when municipalities retain consulting engineers to advise them on what plants to buy for their sewage treatment. The activated sludge process offers two major advantages over the trickling filter, of which there are many still in use. First, the plant itself is considerably smaller for the amount of sewage treated—an obvious advantage in crowded cities. Second, the process, as we learned earlier, treats a lot more sewage per day than the comparable trickling filter.

These primary-secondary systems—whether they include a trickling filter or an activated sludge process—are usually referred to as providing "conventional" treatment. As stated before, they do not completely clean sewage, even when they work at their best, and they are seldom maintained and operated at top efficiency, as we shall soon discuss.

The typical conventional primary-secondary treatment plant, if it happens to be working well, may remove most of the suspended solid materials and organic substances measured by the common test for biological oxygen demand (BOD). But the plant's effluent may still contain more than half the nitrogen and phosphorus found in the raw sewage, and, as we know, these nutrients can seriously contribute to the eutrophication of lakes and watercourses. The conventional plant may also remove substantial numbers of disease-producing microorganisms in sewage. Disinfection of the treated effluent with chlorine re-

* For an illustration and description of how they work, see Appendix E.

duces the numbers a lot more, but even then, no one is certain of how many disease germs may escape the combined processes to be discharged into a stream or lake.

Finally, the conventional systems, especially those with activated sludge processes, present a serious disposal problem other than that caused by the partially purified effluent. The plants pile up ton upon ton of thick, dark gooey sludge, which, incidentally, contains many of the so-called pollutants (including the nutrients) said to be removed by the treatment system. It all adds up to one of the great burdens of modern sewage disposal. Engineers have tried burning, burying, selling, and giving it away—anything to be rid of it, but the endless volume forever plagues sanitation departments and demands new solutions.

Despite all these commendable advances, there is nothing yet in conventional treatment technology that can match the full measure of purification of which nature is capable. Alternatives to the partial measures of conventional treatment methods are still necessary if water pollution is to be controlled by municipalities.

8

Nature Again:
From Ponds to Ditches

After a half-century of developing sewage treatment technology that was good as far as it went, but not good enough, some scientists and engineers began to realize that nature could match the conventional primary-secondary systems with air and sunshine played upon a pond of waste water. Granted the natural process took more time and space, but it required less attention and money to build, operate, and maintain. Furthermore, it could become an attractive part of the landscape. The classic case of how municipal officials became aware of the possibilities came from Fessendon, North Dakota.

In 1928 the town completed a sewage collection system and was about to install a treatment plant when the public treasury ran dry. As a temporary measure, Fessendon's sanitation officials piped the sewage out of town to a large, natural pothole, the low side of which was dammed up so the waste water would make a pond. They would hold it there and hope the smell wouldn't get back to town before the treatment plant money was forthcoming.

Months went by, the money didn't materialize, the pothole slowly filled up, and nothing bad happened. Evaporation and seepage through the new pond's bottom kept the sewage from overflowing. But best of all, there was no smell, not even near the water itself! For that matter, it became a surprisingly attractive pond, considering its origin.

When state health officials analyzed the water, they too were surprised. The raw sewage was actually being treated by the pond itself. Even more startling, the purification was as good as the proposed primary-secondary plant might have provided.

The town never did buy a plant. For years the sewage pond served very well, but the state's sanitary engineers didn't catch up with the remarkable development for twenty years.

In 1948 public sewerage systems were becoming more and more necessary for other North Dakota towns which also couldn't afford them. One town, Maddock, decided to imitate what had worked so well for Fessendon. With some minor design changes, Maddock engineers dug a pond which became a highly effective sewage treatment system. In the next five years a dozen more communities did the same, and their success led the North Dakota State Health Department to recommend the idea widely. This, in turn, prompted the U.S. Public Health Service to study the method with a pilot project at Fayette, Missouri. The research confirmed that here was a good, inexpensive way for small communities to provide the equivalent of primary and secondary treatment for their sewage.

Although the idea that Fessendon stumbled upon was as old as wind and water, the accidental discovery seems to have marked a formal beginning in the United States of an important method for cleaning our growing volumes of sewage. Such bodies of waste water are seldom referred to as sewage ponds anymore, but are called "sewage lagoons," or "stabilization ponds," or "oxidation ponds." Today their action in purifying sewage is much better understood than when they were discovered.*

The sewage lagoon is largely a biological system in which bacteria break down the organic material of the incoming sewage. The residue from the bacteria, rich in nitrogen, phosphorus, potassium, and carbon dioxide, has a high nutritional value for plant growth, which stimulates green or blue-green algae blooms; in the presence of sunlight, these blooms take up the sewage nutrients for their own metabolism. In this same process, the carbon dioxide is converted to oxygen, which is released to the water and in turn stimulates the bacterial action. More such stimulation comes from the oxygen introduced to the pond by the action of the wind on the surface. Thus the bacteria go on and on breaking down the wastes entering the

* For an illustration and description of how a lagoon works, see Appendix E.

pond. When carefully designed for the amount of in-coming sewage, a lagoon can handle the waste in-gredients along with its own wastes (from the bacteria and aquatic plant growth) for a long time.

As in other sewage treatment systems, sludge is formed when solids settle to the bottom of a lagoon, but compared to other methods, where disposal of large sludge accumulations is a continuing difficulty, the lagoon is far less a problem. The material spreads itself over the pond bottom in a thin layer which may take years and years to build up enough to make re-moval necessary. It was once figured out that the sew-age from 1,000 people produced 3,212 cubic feet of sludge per year. If this were spread over the bottom of a 10-acre lagoon (a size that might treat the sewage from 1,000 people), the layer would be only 0.0074 feet thick. At that rate it would take 135 years to build up a foot of sludge—and it still wouldn't be too much of a load for the lagoon.

As lagoons were more widely adopted for the na-tion's smaller communities, the records revealed that they could provide the equivalent of primary-second-ary sewage treatment for considerably less cost than trickling filters or activated sludge plants. In 1968 the Federal Water Pollution Control Administration compared various costs for the three types of systems. The estimated construction cost of a lagoon designed for a million gallons of sewage per day was $112,000, while a trickling filter of the same capacity ran $380,000 and an activated sludge plant cost $428,000. Esti-mated operation and maintenance costs for a 25-year period (in 1968 dollars) showed even greater differences. The lagoon figure was set at $25,750, the trickling filter at $535,000, and the activated sludge at $852,500.

The main disadvantage of a lagoon is a combination of time and space. While air and sunshine are good purifiers, nature's system requires as long as thirty or forty days to treat any given volume of sewage. A pond, therefore, may have to cover many acres to accommodate a town's endless flow of sewage measured in millions of gallons per day. The time and space requirements have been reduced in recent years by the mechanical aeration of lagoons. Large, floating mixers churning the water and exposing it to the oxygen of the air help speed up the natural processes, thereby reducing both the time and space required.

Granting its relatively high demand on land, the lagoon still offers a treatment system of considerable esthetic quality. Instead of a factorylike treatment plant, a town can have an attractive lake or series of lakes treating its sewage. Instead of power demands and possible air pollution from sludge processing, a good lagoon works silently, contributing oxygen to the air and attracting a wide variety of birds and wildlife. In these times when the concern for open space and nature conservation increases, lagoons rather than treatment factories can become more and more attractive.

But don't they smell, the wary citizen is sure to ask. Actually the well-designed, properly operated lagoon may have fewer odor problems than other treatment systems. In colder climates where ice forms in the winter, a lagoon may smell for a short period in the spring as the ice goes out, but otherwise there are no major problems.

The sewage of Belding, Michigan, is treated by a series of lagoons at the edge of town. For some time the town manager requested the Belding police department to run a "sniff patrol." At scheduled times every day a police cruiser stopped at one of several designated points in town, an officer left the car, took a few deep sniffs of the air, and filled out a report. The tests revealed no odors at all except for a matter of a few hours when the ice was breaking up in the lagoons each spring. Even then the odor was only noticeable under certain wind conditions at the side of town nearest the ponds.

In recent years a number of American communities have become interested in a leading European sewage treatment system that incorporates principles from both lagoons and conventional technological treatment methods. The system was developed in the 1950's at Voorschoten, Holland, by Dr. A. Pasveer, a well-known Dutch sanitary engineer. Pasveer called it an "oxidation ditch," although his creation became widely known as the "Pasveer ditch." When introduced in America, the ditches were nicknamed "racetracks" because they were built with endless channels, several feet wide and a few feet deep, cut into the ground in the shape of a typical racetrack, straight on two sides with rounded ends.

Pasveer's ditch, like lagoons, depended on the breakdown of sewage by bacteria, and, like conventional treatment plants, the waste water was enriched with sludge and supplied with plenty of oxygen. In a single ditch Pasveer found he could replace four major elements of an activated sludge process, yet offer a comparable degree of purification. As the ditches were improved, one could treat a given volume of sewage in a day, a thirtieth or fortieth of the time required by a comparable lagoon, but two or three times longer than a good activated sludge system.*

In the mid-1960's the Pasveer ditch was licensed by the government of Holland for sale in the United States by the Lakeside Equipment Corporation of Bartlett, Illinois. In the next seven or eight years, Lakeside installed nearly 350 ditches all over the nation, including one at Fairbanks, Alaska. They were also accepted for sewage treatment in Canada.

An installation to treat a million gallons of sewage a day might require a racetrack 500 feet long and 150 feet wide. The actual ditch might be 40 to 50 feet wide and 4 to 7 feet deep. Low construction cost was a feature of oxidation ditches. Lakeside said that it could underbid the typical activated sludge plant by 20 to 30 percent. Furthermore, the firm claimed operation and maintenance was considerably lower because the system's simplicity gave fewer troubles and required less skill. Finally the ditches, according to the builders, produced less sludge for disposal than the other common treatment systems, except, of course, lagoons.

As the original designers and builders of the Pasveer ditch developed larger units back in Holland, they encountered design, operational, and maintenance problems when their racetracks had to expand enough to

* For an illustration and a description of how an oxidation ditch works, see Appendix E.

treat extra-large volumes of sewage. For a solution, Pasveer came up with a new system, which was called the "Carrousel." It used the same principles as the oxidation ditch, but the Carrousel could handle larger volumes of sewage while occupying considerably less land.*

The new system was also patented in Holland, and many were built and successfully operated in Europe. In the early 1970's the new system was introduced into the United States by a New York licensee, Aerobic Systems, Inc., and it began competing with other systems in the burgeoning business of sewage works from coast to coast.

A look at sanitation facilities around the country reveals that there are still other sewage treatment systems available. For example, there are numerous small "package plants" for sale, relatively compact equipment that might be used to treat the waste water of a small housing development or trailer park. They are usually some variation of the systems we have already discussed. As with the rest, they try, in a controlled, convenient way, to imitate what nature learned to do long ago with soil, plants, air, sunshine, and living organisms, but as usual they don't perform nearly as well.

---

* See Appendix E for illustration and working description.

9

Treatment or Mistreatment

Water pollution control in America has been a long, losing battle covering nearly a century. The generals, the nation's sanitary engineers, have spent billions and billions on weapons, but the battle reports only grow more dismal.

In the late 1950's a Connecticut conservationist in search of an uncontaminated river system in America was advised by the Council of State Governments that none remained to its knowledge. In 1972 the annual report of the federal Council on Environmental Quality reported that while the nation's air was getting cleaner, our water continued to get dirtier. Pollution by sewage from municipalities and industries had not diminished, the council stated, and then explained: "In all types of river basins, the concentration of nutrients is increasing."

How could this happen in a nation where construction of sewage collection and treatment systems since World War II can in some places be compared, in terms of costs, with the building of highways and schools?

Part of the answer is that a 400 percent increase in our waste water load has outrun the development of treatment facilities. Also, less suspect sources of pollution outside the common realm of sewage have added to the burden. The runoff of nutrients from agricultural lands and from urban construction sites has turned out to be a serious contributor to water pollution.

But for the most serious answer as to why water pollution forever increases, one must turn to the most prevalent methods of sewage treatment. As stressed repeatedly in previous chapters, conventional treatment systems are not good enough even when they work at their best. And, as we shall now discuss, very few actually perform the way they should.

These failures in sewage treatment methods are important matters for every concerned citizen to know about as we spend more billions of dollars than ever through the Federal Water Pollution Control Act Amendments of 1972. Under these costly circumstances, the failures of conventional methods demand that we seriously consider alternatives as intelligently and as quickly as possible.

Let's start with an example of what really happens when many communities install conventional sewage treatment plants and conclude they've done their part in fighting water pollution.

In the 1960's Chapel Hill, North Carolina, home of the University of North Carolina, used a $2 million primary-secondary plant to purify its sewage at a purported 85 percent efficiency for BOD reduction, and most townspeople assumed it worked at that level. They were due for a surprise.

Once installed, the plant had been left to run pretty much on its own — as is true with most of the nation's sewage treatment plants. Despite a sizable investment in the sewage works, Chapel Hill had neglected to hire a qualified operator. Such people were hard to find, and they demanded more salary than most towns felt like paying. Instead, Chapel Hill assigned a couple of unqualified men on a part-time basis to watch over the system. Meanwhile the expensive plant churned some 2.5 million gallons of sewage per day through its components and discharged the effluent into a river. What kind of a job was it doing?

Some professors working with Daniel A. Okun, the head of the University of North Carolina's Department of Environmental Sciences and Engineering, decided to find out. With the town's permission they analyzed the plant and found it was not operating at its rated 85 percent efficiency; it was down around 60 percent.

Okun and his colleagues then asked if they might operate the sewage plant, and the town agreed. Soon Chapel Hill had an operations crew unlike any other in the world, headed by a professor and manned by a staff of environmental specialists. They did marvels for the treatment plant, simply by raising its efficiency at BOD reduction to its rated value. They thereby put the plant into an operational class relatively unique in America, even though the best could not be called good enough to control pollution.

As such plants were developed and installed over the years, most were sold with the understanding that they could remove a good percentage of sewage ingredients responsible for water pollution, even though the cleaning job was not enough. And they could keep the promise, if properly operated and maintained. But during the sale of an expensive system, the equally expensive "if" was seldom emphasized enough, and in the end most plants never kept up to the performance measures built in by the manufacturers.

In the 1971 Senate hearings on water pollution and waste water treatment, Daniel Okun testified that, ". . . except for a few unique treatment facilities, operated by well-qualified and dedicated professionals, most of the treatment plants in the United States are operating far below the efficiency for which they were designed and fail to even meet the very limited objectives set for them by the investment of relatively large sums of money."

Okun's remarks were echoed in the same year by Russell and Gordon Culp, father and son who are highly respected in the sanitary engineering field. In a book on the subject they said: "It is sad but true that operation of the great majority of existing secondary waste water treatment plants is quite erratic and often badly neglected. Once built, many plants are then ignored or their existence almost forgotten. In many, many plants virtually the entire investment in sewage treatment facilities is lost for all practical purposes, because the plants were never really operated, and the potential benefits of pollution abatement were never fully realized." As a result, the Culps concluded, the public gets a lot less for its sewage money than people think.

Far too many of the nation's multimillion dollar treatment plants are run by low paid, unskilled, poorly educated individuals who need to be replaced by persons with a basic knowledge of chemistry, biology, microbiology, bacteriology, physics, and mathematics. A qualified operator, who demands a relatively high salary, is essential to perform, record, and analyze laboratory tests to determine if his plant is performing efficiently. If he can't do this, the plant's troubles are certain to increase, often without his even recognizing the difficulties, to say nothing of his knowing the remedies. Nor does the paying public know of the problems—until local water pollution tells the story.

A conventional sewage treatment plant, depending on biological (bacterial) processes, is a potential package of problems. Efficient operation requires the well-being of billions of live bacteria—or "bugs," as many operators call them. Conditions must be just right at the treatment plant if the "bugs" are to function at their best. If not, the plant's effectiveness suffers, often drastically.

Above all, the operator concerned for his plant's biological processes must constantly monitor the incoming sewage for substances harmful to the crucial bacteria. Especially he must guard against the influx of toxic materials that can wipe out the biological heart of his plant, sometimes in minutes. Indeed, the diligent, well-qualified operator will try to prevent such disasters in advance by going to the sources of the sewage, especially to the system's industrial users, to make certain that detrimental substances never get into the system in the first place.

The sanitation districts of Los Angeles County, which have many sewage plants, provide the essential kind of operation and maintenance programs missing in most communities. The districts run extensive laboratory testing programs with staffs of skilled technicians who constantly monitor the ongoing processes. To prevent toxic substances from damaging their biological systems the technicians are virtually detectives. They run down every possible source of toxic ingredients to make sure they're not discharged into the public sewer system.

One time they were stumped when, one after another, the districts' sewage plants were being knocked out by a toxic waste from an electroplating process. While the sanitary detectives were trying but failing to find the culprits responsible for the failure at one sewage plant, another would suddenly suffer the same fate. The mysterious source was finally caught. It was a small electroplating plant whose managers were avoiding the cost of properly discarding their toxic waste. They knew they couldn't get away with dumping it in their own sewage line, so they let the waste accumulate in barrels and occasionally trucked it through the dark of night to some point in Los Angeles County, opened a manhole on a lonely street, and dumped the potent residue into the local sewerage system. In short order the surprised operator of the sewage plant on that system discovered his purification process had been dealt a mysterious death blow; the county was minus a treatment plant until it could be cleaned and restored to service.

But the capability of Los Angeles County remains relatively unique, even though wide efforts have been made to upgrade the operation and maintenance of America's sewage treatment. In 1963, for example, the federal government distributed recommendations for a minimum operations and maintenance program for municipal treatment plants. The guidelines were developed by a conference of state sanitary engineers cooperating with the Federal Water Pollution Control Administration. But were they followed? One authoritative study indicated that relatively few plants applied the guidelines. Of 69 plants surveyed around the country, 59 were not following all the recommendations in 158 instances.

The investigators, who came from the U.S. General Accounting Office, turned up the case of an untrained operator who continued reporting important flow data, even though his flowmeter had long ceased to work. Another such employee was recording the results of laboratory tests even though a thick coat of dust on his equipment revealed that tests had not been conducted for a long time. In another case an inexperienced operator was unaware that raw sewage rather than treated effluent was pouring from his plant into a nearby stream, because he had failed to recognize that a local poultry firm was ruining the purification process by dumping grease and feathers into the public sewer. Maintenance had failed so badly in another community that the treatment plant was near self-destruction. Finally, the investigators were hardly surprised to find a town where the people were oblivious to the fact that their raw sewage was flowing virtually unchanged from the inlet to the outlet of their expensive but largely unattended treatment plant. Such cases had a lot to say about the failure of conventional sewage treatment to prevent the incessant increase of water pollution.

Anyone assessing conventional sewage treatment must know about "bypassing," a technical term for what usually happens when a plant is seriously overloaded or breaks down. Since the inflow of sewage never stops, it has to go somewhere whether the system is working or not. When something goes wrong, the waste water may be shunted around the plant or simply allowed to run on through with little or no purification. In some of the foregoing examples of poor operations we have already alluded to bypassing without using the word. It happens much too often in America, and seriously large volumes of raw or poorly treated sewage are discharged by municipalities.

A main cause goes back to the time-worn fact that many communities have "combined" sewage collection systems, which means they accept storm water from urban streets as well as domestic sewage from homes and industries. In fair weather the normal waste water load may be readily treated by the sewage plant, but then a heavy rain can suddenly overload the purification system, and a large measure of raw sewage passes around or through the plant inadequately treated.

Chicago has long been troubled by this problem. Only an inch of rain gathered in the vast paved acres of Cook County causes torrents of water to overload the combined sewerage system. Despite all the city's treatment capacity, the crush of storm water forces a lot of raw sewage to spill out of the system, and it flows directly to the Sanitation and Ship Canal, to the Illinois River, and on to the Mississippi.

Another cause of bypassing may be found where poorly constructed sewer pipes, which literally leak in reverse, pick up a great deal of groundwater. The added volume swells the sewage load beyond the treatment plant's capacity, and a large part of the raw waste is bypassed to the receiving stream.

Bypassing can suddenly become a necessity when mechanical failures simply force a plant, or its allied equipment, to shut down. Thus it's conceivable that with one fairly simple electrical or mechanical failure, a treatment plant can be forced to drop from a relatively high degree of purification to zero in a matter of minutes.

Of course, there are solutions for these problems, but they're often expensive. The most costly would be to have enough extra treatment plants, like spare tires, to compensate for breakdowns and to handle even the greatest peak load—which, most of the time, could mean that a lot of costly equipment would be working far below capacity. For a less expensive but still costly solution, large storage basins may be built and held ready to collect bypassed sewage until the treatment plant can handle the load. The empty basins may occupy a lot of land while waiting to receive hundreds of thousands or millions of gallons of sewage in an emergency.

Chicago is trying to solve its serious bypassing

problems by carving huge underground tunnels from the rock below the city. Excessive storm water from the streets will be caught in the cavernous tubes and held until it can be treated properly.

All these on-going problems of operation and maintenance should be, but are seldom, considered carefully enough by municipal officials and citizens when they adopt a sewage treatment system for their waste water. These are problems, along with all the others, that once more emphasize why it's important that concerned citizens thoroughly consider alternative methods of sewage treatment—and if they do, nature's remarkable purification systems may again turn out to have a lot to offer.

Lagoons, for example, have more potential than most conventional treatment systems for successfully reducing the problems that cause bypassing. Take, for instance, the difficulties of peak loads caused by a sudden volume of storm water runoff. A well-designed lagoon or series of lagoons can allow for enough capacity to accommodate such peak loads. The water level simply rises, and in due time, the lagoon treats the additional sewage along with its routine load.

Also, the well-designed lagoon is, in a sense, a more resilient system when it receives the kind of toxic substances that can knock out a biological plant, like a trickling filter or activated sludge process. First of all, lagoons, which inherently take more time for treatment, are slower to react adversely. If a toxic material enters a lagoon, its effect may spread through the large volume of water at a relatively slow rate compared to the sudden death the substance may deal to the crucial bacteria in the comparatively confined spaces of a conventional treatment plant. The slow reaction of a lagoon gives the operator time to discover and correct the problem before very much of the treatment capability is destroyed.

Well-designed land treatment systems, perhaps working in conjunction with lagoons, are also comparatively trouble free. As we shall find out, this factor has led certain industries with extremely potent waste water to treat it through the use of soil and plants in land systems.

But these alternatives are not being considered enough by the sanitary authorities who should be most concerned about the failure of the conventional factorylike plants which they continue to recommend for the nation's water pollution control efforts. When it comes to these plants not doing the control job, even at their rated best, sanitary engineers still are not prone to look at alternatives; they are likely to recommend more of the same kind of technology. They claim that primary-secondary plants need more stages of mechanical, chemical, or biological purification. They speak about them as tertiary (or advanced) treatment, and as if it offers the final answer to our long, wearisome difficulties with sewage.

Those who take a careful, critical look at the contention find it isn't necessarily so.

10

Tertiary Trials
and Tribulations

The federal clean water bill, which became public law in October 1972, states in the beginning, ". . . it is the national goal that the discharge of pollutants into the navigable waters be eliminated by 1985." If this goal is attained it will require that we soon develop the capabilities for purifying sewage considerably beyond today's conventional, primary-secondary effluent. How do we do it?

Most established authorities in the sanitary engineering field claim that tertiary treatment is the answer. However, as the 1972 law was passed, engineers had made little headway with tertiary treatment works. The municipal examples were only two, and they were very far apart. One, in Windhoek, South Africa, was purifying a relatively small amount of sewage well enough to mix the effluent with the community's drinking water supply. The other was at beautiful, mile-high Lake Tahoe on the California-Nevada border. It had been in operation for seven years, had become the showplace for many proponents of tertiary treatment, and was as well the subject of controversy among sanitary engineers as to its effectiveness.

Lake Tahoe is one of the three deepest, clearest lakes on earth, rivaled only by Lake Baikal in Russia and Crater Lake in Oregon. In the late 1950's a population and tourist boom on Tahoe's south shore tremendously increased the volume of sewage at the town of South Lake Tahoe. A primary-secondary treatment plant was hastily constructed, but the effluent disposal was an extremely difficult problem in the mountain-ringed Tahoe basin where all but one of the sixty streams flowed into the beautiful lake. Local conservationists fought hard and successfully to insure that not a drop of the sewage water would touch Tahoe's deep, blue waters.

The local sanitation board tried spraying the treated effluent on a hillside, hoping the waste residue would disappear into the ground. But that solution was poorly designed, indeed not designed at all for a workable land treatment system. The weather, the soil, the slope of the land, and the small available acreage prevented all of the large supply of effluent from infiltrating into the earth to be purified as it will in a properly designed land system. The runoff threatened to get to the lake, and this brought cries from the conservationists. Then one Labor Day weekend the treatment plant failed, about two million gallons of raw sewage had to be bypassed around the system, and some of it did reach the lake. This raised the public's hackles more than ever, and South Lake Tahoe decided, one way or another, to find a means of completely purifying the town's sewage so it couldn't possibly pollute the lake.

Two consulting engineers, Harlan Moyer and Russell Culp, were given the difficult job. The two consultants decided they could meet the need technologically, not necessarily with miraculous new discoveries but with existing equipment from various industries. Moyer and Culp started with a 25-gallon-a-day pilot model developed in the sewage plant garage. Then a full-scale tertiary plant was built with local tax money and federal funds. When the plant was completed in 1965, raw sewage flowed into the system, and water with large percentages of the pollutants removed flowed out. For example, 94 percent of the phosphorus was removed and from 50 to 98 percent of the nitrogen, according to the designers.

The new tertiary part of the treatment plant was a large, complicated maze of technology that carried the secondary effluent through five steps of purification. The first was a chemical treatment process that removed phosphorus with lime added to the effluent. Next, the effluent was pumped to a tall "ammonia stripping" tower, and as it flowed down through a maze of baffles, nitrogen was removed, but in time the performance of the tower was troubled by low temperatures, so it wasn't always efficient.

The effluent was then led through some new and sophisticated "mixed multi-media filters" made with coal, garnet, and sand, a combination that was designed to require comparatively little flushing out, as it removed very fine particulate matter from the water. The fourth stage drew on a process used in the sugar industry where carbon was employed to remove color from raw sugar. The carbon method was adopted at Lake Tahoe to decolorize the effluent so it could leave the plant as clear water. In a final stage the effluent was disinfected with chlorine to kill any disease organisms that might have survived all the chemicals, pumps, filters, mixers, and other processing units.

A fine gray ash also came from the plant, residue from the ordinary sludge produced by the conventional primary and secondary treatment stages. At Tahoe it was run through thickening equipment, a centrifuge, and a "biological sludge furnace," which required anti-air pollution devices to prevent dirtying the mountain air. The end product was an inert dry, light-gray ash that had to be disposed of.

When complete, the cost of the Tahoe tertiary plant was reported to be about the same as a primary-secondary system. In other words, it about doubled the cost of the community's sewage treatment system.

While the engineers at Tahoe frequently drank their tertiary effluent to prove it was safe for human consumption, the protectors of Lake Tahoe remained wary. They kept up their opposition to having any sewage plant discharge get to the lake, even if it was fit for drinking water. "We could have distilled the effluent down to plain, pure H_2O," said Moyer, "and they still would have cried, 'Get it out of here! It's still sewage!'"

At the same time, the Tahoe conservationists were

waging a successful battle to have the sewage effluent from all the communities around the lake piped out of the Tahoe Basin. When this was finally ordered, South Lake Tahoe was forced to pump its tertiary effluent twenty-eight miles over a mountain pass to where the water became the sole supply for a new manmade lake, Indian Creek Reservoir. Fish thrived in the reservoir, and it was approved by the State of California for recreation, including swimming.

From inlet to outlet the South Tahoe treatment system was a collection of "unit processes." Each process served as a specific cleaning unit designed to purify sewage one unit at a time until enough of the wastes were gone for the water to be drinkable.

This line of unit processes was the basic idea behind most tertiary systems being planned and built at the start of the 1970's, although the units were not always the same as at Tahoe. Design engineers could often choose from several possible processes to remove a specific ingredient, such as phosphorus or nitrogen. They were often expensive and difficult to maintain.

As tertiary treatment plants were being discussed and tested in pilot projects, a number of disadvantages became obvious.

They would be large, complicated, and expensive systems subject to the shortcomings of conventional treatment technology—only a lot more so. At a time when sanitation departments were failing to keep primary-secondary plants working well, many communities were being advised to multiply their operations and maintenance problems by adding more and more sophisticated tertiary processes.

The advanced systems, to the dismay of some ecologists, would use up natural resources, mainly such chemicals as lime and carbon, to retrieve another natural resource, water. For example, one of the largest advanced systems, now under construction in the Washington, D.C., area, has plans for purifying 309 million gallons a day so that the effluent will not pollute the Potomac River. An expansion of the conventional Blue Plains treatment plant, only a few miles from the heart of the nation's capital, it will become a full-scale tertiary system by the mid-1970's. It is estimated that this huge plant will devour over 460 tons per day of seven different chemicals, and the daily cost (at 1972 prices) will total over $23,000.

The main purpose of such advanced waste water technology is simply to turn raw sewage into water pure enough to dump with impunity — but what happens to the pollutants? They don't vanish. Actually, most will end up in the sludge. What's more, the sludge will now contain more than the wastes removed from the sewage; it will hold the residue of the processing chemicals — and they are likely to make disposal problems all the more difficult.

This is a concern at Blue Plains where the disposal of primary-secondary sludge has already become a difficult, expensive problem. With the coming of tertiary processes, sanitary engineers wonder whether the sludge residue of treatment chemicals might substantially affect groundwater supplies or the growth of plants if the sludge is disposed of on land, which seems like the best practice. The potential troubles are being researched by the U.S. Department of Agriculture at Beltsville, Maryland, by a team of micro-biologists, agronomists, soil scientists, plant physiologists, pathologists, and others.

When the no-pollution goal was put into law, sanitary technologists had a lot of plans and ideas for completely separating the 99-plus percent clean water of sewage from the fraction percent contaminants, but they still have a long way to go before sewage factories will be able to thoroughly clean appreciable volumes of America's waste water.

Can such mills work on such a grand scale extending indefinitely into the future? Will their environmental impact—possible air pollution, energy consumption, and endless demands on mineral resources—be as bad as the water pollution they aim to end? Will they be worth the cost, especially for operating and maintaining their expensive processes?

For some places the answer to all these questions may be "yes," but for a large majority it may be "no!"

Such questions have sent numerous scientists, engineers, and government officials looking for alternatives to tertiary treatment by technology. The best alternative, some have concluded, is to go back to the land with our waste water. This could offer us a viable form of tertiary treatment with which to meet the national goal of 1985—plus all the benefits of living plants beautifying open space that would add to not subtract from natural resources, improve not jeopardize the air, enhance not detract from natural beauty, and cost less in the long run.

PART 4

The Effluent
Treatment of Land

11

When Land Treatment
Became Tertiary Treatment

In 1965 Professor George A. Whetstone of Texas
Technological College published a 150-page bibliography on the reclamation of sewage effluent, which
was introduced saying: "That reuse of water through
many cycles will be routine practice in fifty years
seems so evident after a diligent survey of the technical literature, that the writer toyed facetiously with
assigning to this study some such title as, 'The 21st
Century: An Effluent Society.'"

The Texan's bibliography was one of many con-
tributions in recent times by scientists, engineers, and
others revealing that while municipal sanitation had
become dominated by dilution and insufficient treatment methods, the remarkable purification powers of
soil and plants had not been forgotten. From the

99

1930's to the present, one could still find land treatment at work in America, not for raw sewage, but as a form of "tertiary" treatment to complete the purification of sewage effluent that would otherwise contribute to the nation's growing water pollution. In the early 1960's the effluent from 401 municipal sewage treatment plants in the United States was receiving additional land treatment. In many of these instances, it is safe to say, the municipalities were successfully providing tertiary treatment during the exact period when sanitation engineers were trying with far less than complete success to perfect advanced waste water treatment systems using mechanical, chemical, and biological means.

One of the most successful examples of tertiary treatment by land has been in operation since the 1930's at Lubbock, Texas, where a farmer, J. Frank Gray, has performed two big jobs simultaneously. While managing and operating one of the best farms on the Texas High Plains, he has also treated the sewage effluent from the adjacent city of Lubbock. Both Gray and his fellow citizens reap the benefits. The farmer enjoys outstanding crops and soil improvement, and the citizens have water pollution control and an abundance of fresh water for community beautification and recreation.

In Lubbock in the 1930's a new primary-secondary sewage treatment plant with a trickling filter was installed at the edge of town. While most such communities were simply dumping secondary effluent in the nearest river, this wasn't possible at Lubbock without ruining a local lake which was a recreational treasure in that arid countryside.

However, some Lubbock officials were aware that the potential pollutants were also potential agricultural resources, so the effluent could be simultaneously used for irrigation and fertilizer. And they understood that, during irrigation, nature's astonishing purification processes would finish cleaning the effluent. With this in mind they decided to try the old method of land treatment.

After they had acquired about one hundred acres of farmland for irrigation, the city officials were offered the use of an adjacent 126-acre horse farm owned by a local physician, Dr. Fred Standefer. Crops in both farms soon thrived better than others in the vicinity without any public health or nuisance problems. Recognizing the potentials of the effluent, Dr. Standefer made the city an offer. In return for the community's waste water, he would dispose of it indefinitely. He would expand his farming operation and lease the city farmland for his own use. Lubbock agreed, and Dr. Standefer was in business with a land treatment farm. In 1936 he hired Frank Gray as his manager. Gray, who was just completing his college education in agriculture, soon became the doctor's partner, and when the latter died, the young man became the principal farm operator under the agreement with the city.

Meanwhile Lubbock, like many cities, expanded rapidly — and so did the volume of sewage effluent. To handle it, Gray frequently added land to his farm. With the city's help he also developed storage ponds to cope with the usual variations in the growing volume of effluent. By the 1970's, when Lubbock's population had reached nearly 170,000, he had some 2,600 acres under irrigation and was supplying effluent to a neighbor for another 1,600 acres.

Over the years Gray produced various crops, including wheat, barley, oats, and rye. He raised livestock on irrigated pasturelands. For several years he even maintained a grade-A dairy farm, with the cows feeding on sewage-irrigated crops, and the milk never failed to pass the state's demanding health tests. Meanwhile, the sewage resources helped the farmer build up some of the richest soil on the High Plains.

The best part of the Gray farm story is still in progress. As the land was irrigated, and the water filtered into the earth, billions of gallons collected in a huge natural underground storage area. This contrasted sharply to what was generally true on the plains, where the water table was dropping ominously because of the persistent demand from conventional irrigation wells. Tests on the Gray farm in recent years revealed that the table had risen some fifty feet since the 1930's.

Furthermore, the vast underground reservoir tested virtually as pure as the pristine waters that must have been under Texas in ages past. The only negative part of the water tests showed a high nitrate content. This probably resulted from soil oversaturation when the ever-increasing volume of waste water outran the farmer's land expansions. While the nitrate content would eliminate the supply as drinking water, it was not serious enough to prevent use of the water for most other purposes, like recreation, possibly to include swimming.

In the late 1960's Lubbock city planners proposed bringing the purified waste water back into town for everyone's benefit. The plan involved Yellow House Canyon, site of dried-up Yellow House Creek and the only topographic relief in the flat country where Lubbock was built. The canyon, which cuts across the heart of the town, might have remained the prettiest part of the city, but it had been allowed to become the community's poverty row and dumping yards. The planners proposed restoring the canyon so that it could again become Lubbock's most beautiful natural area. It was possible because of the huge water supply under the nearby Gray farm.

When plans were ready, some $20 million were sought from federal sources and from Lubbock voters who would have to approve a major bond issue for the canyon project. The main campaign tool, a homemade slide film with a tape-recorded narration by a local radio announcer, was shown dozens of times to thousands of people. Also, students of architecture at a local college made a magnificent model of the proposed canyon restoration for display to hundreds of visitors at city hall.

The plans called for turning the canyon into a green belt extending through the city. It would include eight small lakes, six within the city, filled with water drawn from twenty-seven new wells on Frank Gray's farm. The water would also serve to irrigate trees, shrubs, flowers, and grass on the canyon's sides, which would provide for all kinds of recreation—picnics, nature walks, playgrounds, and for simply sitting to enjoy the surroundings. The lakes themselves would be used for boating and fishing, and swimming might eventually be permitted.

The voters approved the bond issue for $2.8 million by a three-to-one majority. The money was more than matched by several federal funds, so the city had $7.6 million to begin the unusual project. It was started by purchasing land, building the six lakes, and drilling the wells on Gray's farm to draw some 2.5 million gallons of water per day to be piped into the city. During the 1970's Lubbock plans to complete the conversion of Yellow House Canyon from the most desecrated to the most beautiful area in town.

Thus the story of Lubbock's sewage is truly one of turning waste into wealth by working within nature's wheel of life. It has been the most successful of many examples of effluent irrigation serving as tertiary treatment in the United States. However, practically all such land systems have been in the arid and semiarid states of the West, where water shortages make water pollution an extremely serious matter and where farms badly need irrigation water.

The most concentrated use of effluent for irrigation has been in California, where growing communities before and after World War II were seriously threatened by water shortages and pollution of dwindling water resources. In 1951 forty-two towns and cities in California's central valley were irrigating crops with treated effluent. The largest, most publicized project was at Fresno, population 112,000. For many years some 18 million gallons of raw sewage a day were piped five miles out of the city to a primary plant. The treated effluent was then delivered to storage ponds covering 305 acres of a 1,292-acre municipal farm, where it was applied to crops, including 65 acres of alfalfa, 90 of Sudan grass, 190 of Kaffir corn, and 600 of pasture for the farm's 600 head of Hereford cattle. The Fresno sewage treatment system had the distinction of turning a profit for several years. In 1949, for example, $9,346 was returned to the city treasury.

In the 1940's and '50's numerous other successful irrigation projects were in operation in the water-short, pollution-threatened West and Southwest. Tucson, Arizona, typified most of them. About 3 million gallons per day of primary effluent irrigated a city-owned farm run by six men who raised oats, barley, and ensilage with profits of $3,000 to $5,000 a year.

The fact that land treatment was confined to areas with long growing seasons and dry climates often led to the assumption that the method was unusable in wetter, colder regions. However, this was proven false by European projects. As had happened seventy-five years earlier, American observers visiting Europe in the 1950's found land treatment in use and being improved.

In Bavaria, for example, the government had appropriated a million dollars for research facilities and pilot projects which had led to twenty-eight land treatment installations in a few years. Meanwhile, Bavarian legislation had ruled out any subsidy for sewage treatment until a thorough study had explored the possibilities of land treatment.

By this time the European experts had developed methods of effluent irrigation adapted to colder, wetter climates, soils, topography, types of crops, groundwater tables, and local farming practices. The techniques depended largely on careful applications of effluent intended to maximize the annual period when irrigation could be practiced successfully. In off seasons lagoons collected and held effluent until the weather made irrigation possible.

One of the most important techniques was developed in the 1930's when European irrigators sprayed effluent on their land, instead of relying on the old methods of flooding fields. By the 1950's spray methods were widely adopted in the Old World, but the limited land treatment in the United States still depended on irrigation.

A report from Poland, for instance, told of more than twenty municipalities and many independent industrial works discharging effluent from movable pipes that delivered the waste water to the land as would a fire hose. In other cases it was sprinkled upon the earth by forcing the effluent through small holes in perforated pipes. Or the waste water was thrown in wide circles by revolving spray nozzles, one of which could cover as much as 2.5 acres. This latter method became the most widely adopted spray technique. The success of effluent spraying in general was attributed to several important advantages:

- The nutrient-laden waters could be made to fall

gently on crops like rain, and the land received a uniform dose regardless of topography. In fact, in Europe the sprayed effluent became known as "artificial rain."

• The irrigator had far greater control over how much and exactly where he delivered effluent. He had the important advantage of being able to apply small but frequent doses to crops, as opposed to the large, less controllable amounts in conventional flood irrigation.

• With portable spray equipment a single operator could treat a lot more land with much less time and trouble.

• Without the ditches and dikes required by flood irrigation, tractors and other farm equipment could work the sprayed land much faster and more conveniently.

• The loss of effluent by infiltration into the earthen bottoms of irrigation ditches was eliminated by spraying, an important consideration for farmers concerned about the fertilizer value of effluent.

• Finally, the spray technique resulted in higher crop yields.

Thus the advent of effluent spraying on land treatment farms offered opportunities to improve management methods, which had always been the key to success in sewage irrigation projects.

In a 1956 report on European practices, Bernard P. Skulte, a consulting engineer from Boston, concluded: "Spraying . . . is hygienically and economically the best means of sewage utilization. Undisturbed natural growth of plants, such as grasses, also promote the development of good soil structure with optimum utilization and percolation rates. The spray irrigation system has been in continuous service in Western Europe for more than two decades, and has been entirely satisfactory from every viewpoint. Crop yields have been increased markedly through well-managed irrigation."

Skulte also reported on some other innovations that would reduce costs and allow for improved management methods. Light, flexible plastic pipe was being used to make spray rigs easier and faster to move. Delitsch, Germany, employed special pipe with a porous top laid about two feet below the ground surface in a fishbone pattern. When effluent was forced through and up out of the porous pipe, it impregnated the soil around the crops' roots. Skulte stated: "This system enables greater root penetration and expansion, and correspondingly increases the yield of crops."

Spray irrigation opened new possibilities for another important land treatment idea: forest irrigation. It was recognized for a long time that trees and the forest floor could absorb a lot of water. In 1890 an English book on sewage disposal recommended planting osiers (a variety of willows) to increase the absorptive quality of land to be treated with sewage. It was also widely recognized that the undisturbed forest duff (natural litter) could receive great volumes of water, even in freezing weather. In the 1940's an American investigator showed that the duff could accept three times the water absorbed by a cleared, eroded area. Another researcher found that rainwater runoff from cultivated watersheds exceeded that from woodlands and pastures by more than ten times. Such data was certainly of interest to the proponents of land treatment, for it pointed the way to dealing with large volumes of waste water.

When these innovations found their way to America, they were first used in certain industries to treat some of the country's most potent waste waters — those that were causing serious water pollution as well as calamities for conventional sewage treatment systems. Land treatment in these instances proved its capabilities under difficult circumstances in all parts of the country.

12

The Salvation
of the Biggest Polluters

Some of the nation's worst polluters of rivers, streams, and lakes had to do something about their waste waters long ago or suffer more public wrath than business could bear. This was especially true of large food processing factories and dairy plants.

Their waste waters had an extremely high biological oxygen demand (BOD). While common domestic sewage had a BOD of around 200, a food processing plant's waste water might test three, four, five, or more times greater. Thus a single food or dairy plant could be a more serious polluter of local streams than all the rest of a town put together.

Furthermore, many food industries had a pattern of intermittent peak discharges that were anticipated with horror at municipal sewage plants. Months might go by with little or no flow, but then, during the harvest and processing period, the local sewage plant was literally drowned in industrial waste water more potent than the sanitation department could treat well even in normal volumes. The waste swept through the plant largely untreated and delivered a disastrous blow to the fresh water stream receiving the deluge.

The only fair answer was for the industries to treat their own sewage. But the capital costs of typical sewage treatment plants were too high for most comptrollers' budgets. Not only were the conventional systems expensive to buy; their operation, under the wildly varying waste loads of many firms, would also have been difficult and costly. When company engineers looked for alternatives, they often settled on land treatment. At first they tried conventional flood irrigation, but soon the European spraying methods were widely adopted and improved on.

By the 1950's more than twenty United States canning and frozen food firms were spraying their extremely potent liquid wastes on the land. As the country became more conscious of water pollution, the number of industrial spray irrigators increased even more. The installations were usually specially designed and operated with considerable study, thus the science and art of land treatment was rapidly improved under many climatic conditions.

At Dayton, Washington, for example, the Touchet River was seriously jeopardized in the mid-1950's by the large Green Giant cannery, which operated only a month a year. From about June 15 to July 15 pea-canning processes produced nearly 1.5 million gallons of waste water per day with the extremely high BOD rating of 1,900. When it was mixed with the city's less potent sewage the resulting BOD was still 1,200. At best the local treatment plant could reduce it to no less than 600, which was still three times stronger than the community's untreated raw sewage. The bad situation was even more critical because in the summer the river's normal volume dropped to about a quarter of its full flow. Dilution, therefore, was not sufficient to help reduce the polluting power of the cannery wastes.

In 1956 Green Giant installed a spray irrigation system on one hundred acres mostly planted with asparagus. That summer the asparagus thrived, Dayton's municipal treatment plant was relieved of its annual overburden, and the Touchet River was protected.

About then another such pollution threat was abated at Franklin, Virginia, when engineers at the Union Bag-Camp Paper Corporation turned farmers. The firm's expanding production of paper increased its volume of liquid wastes, which had a BOD over 1,500, so that a lagoon system, which had reduced the potency to around 200, was now much less effective. Discharge from the lagoons to a local stream threatened serious water pollution.

Company engineers decided to spray irrigate. From April through November nearly 200 million gallons of the lagoon water were sprayed on about 100 acres of crops and 50 acres of woodland. The crops were envied by all the farmers in Isle of Wight County, Virginia. Irrigated corn yielded an average 135 bushels per acre compared to nearby averages as low as 15, with none higher than 80. The county's average peanut production of 1,550 pounds per acre was outdone, with acres yielding more than 2,000 pounds. Between growing seasons the irrigators disposed of large volumes of waste with forest sprays. The woodland soil and trees absorbed more water than open fields, without any evident harm to the forest growth. The total price of the unusual treatment system including land and equipment was about $75,000, plus a daily operating cost of $25. It was an amazingly inexpensive tertiary treatment system that brought with it a bumper crop of quality corn and peanuts.

The classic of all the early industrial land treatment projects occurred at Seabrook Farms Company, a pioneer in growing and quick-freezing vegetables. The company's large factory and vegetable fields are on some twenty thousand acres of land in southern New Jersey. Around World War II, Seabrook faced a multimillion dollar damage suit when waste water from vegetable washings polluted a nearby creek that ran past an amusement park which then sought legal relief. In 1944 the company was ordered to install sewage treatment facilities.

This put Seabrook in a difficult dilemma. Its 5 to 10 million gallons of processing water a day meant the company's waste water volume was comparable to that of many towns and cities. To buy and operate a municipal treatment plant would tilt the firm's profits into the loss column, for such a system cost about $7,750,000. These pressures soon led Seabrook to the relatively new idea of land treatment, and the company retained a Johns Hopkins University climatologist,

Dr. Charles W. Thornthwaite, to work with a company agricultural specialist, Donald Parmlee, to design a land system.

First, Thornthwaite and Parmlee set up a spray irrigation test in a field of clover, but in a few hours, after they had applied the equivalent of only two inches of water, the soil accepted no more, pools formed, and the remainder ran off to the troubled creek.

They then sprayed some nearby forest land and found the soil so porous that it accepted the equivalent of four feet of water in sixteen hours, and all of it vanished into the earth. If this would work continuously in the large forest tracts owned by the food company, Thornthwaite and Parmlee believed they had the answer to Seabrook's difficult waste water problems.

In 1950 Seabrook's forest spraying began on a grand scale. The water, with its major solid materials screened out, was distributed to various woodland areas by open canals. It was then pumped through whirling spray rigs, each set up to cover about 180 wooded acres. Millions and millions of gallons were successfully applied that year and henceforth. Today the sprays are still delivering the waste water to the Jersey forest. Meanwhile the Seabrook installation has become one of the most studied irrigation projects in history.

Even as the original spray sites were selected, twenty-nine sampling wells were drilled around the forest floor, and the natural ground water was analyzed chemically and biologically. To everyone's surprise the natural groundwater didn't meet state standards for potable water because the BOD rating was slightly high. But then the researchers had a bigger surprise. As millions of gallons of waste water soaked through the earth and the water table rose fast, the unpotable ground water actually improved—in fact, it soon met the state requirements for potable water. Only one conclusion could be drawn. The soil was purifying the large volumes of waste water which then diluted the poor quality groundwater enough to make the end product acceptable as drinking water.

After several years of testing, a report on the project said, ". . . it is clear that the woods area was a most efficient disposal plant, rapidly converting polluted industrial effluent into potable water by means of percolation through the ground." At the time, 180 acres of woodland were absorbing about 50 million gallons of waste water a week through the soil, evaporation, and transpiration. In the annual eight-month irrigation period, the land, had it not absorbed the liquid, would have been covered by waste water fifty feet deep.

While the new system protected the natural water supplies, the designers eventually worried about the way the powerful sprays were badly damaging the area's natural growth. The force of the water broke limbs and tore away leaves. The smallest trees died. Even large chestnut oaks were in trouble. Blueberry and huckleberry bushes common to the Jersey woodlands were obviously disappearing. The irrigators feared that the loss of groundcover would allow serious soil erosion.

But nature then alleviated the worries by compensating for the new climate imposed on the woodlands. In the battered spray areas, the native blueberries and huckleberries gave way to pigweed, fireweed, horseweed, and poke. Some of the weeds grew as much as twelve feet high with stalks over two inches thick. Seven years after the sprays had been turned on, most of the native vegetation had disappeared completely and had been replaced by moisture-loving herbs and shrubs.

Instead of eroding, the forest floor had adapted itself to the new environment. Its organic mat had thickened in the sprayed areas. While the mat in unsprayed parts of the forest averaged 3.2 inches, it had increased to 5.3 inches where sprayed. The added mat came from the litter of the heavy, new herbaceous growth and from organic ingredients in the waste water. Also this top layer of the sprayed soil had more nitrogen and phosphorus than previously. And it contained more earthworms than before, which helped create a mulllike humus. In summary, the thicker forest floor had become richer and even more water-absorbent than before.

Today a stranger unaware of the forest's history could be extremely puzzled walking through the Seabrook woodlands when the sprays are turned off. Among the native oaks and hickories he would encounter large circular areas, some two hundred feet in diameter, that would make him think of another land, a jungle of sorts where heavy endless rains had produced growth in sharp contrast to the New Jersey surroundings.

This kind of impact, in today's environmentally

concerned world, might cause some people to shudder. But careful consideration of the forced change in ecology — compared to the distasteful kind that results when misplaced resources create water pollution — may lessen the concern. The New Jersey project is an instance of nature and man working together to come up with a revised ecology mutually acceptable to both parties.

The Seabrook Farms system became a prototype for many more in various kinds of industry. Thornthwaite established a consulting firm which was later joined by Parmlee, and they and other consultants designed many land treatment systems for industries having to discharge large volumes of badly polluted waste waters.

While most imitators of Seabrook Farms succeeded, some failed. An expert look at the failures often revealed that while the spray hardware had been adapted, the designers had neglected to consider how local soils and vegetation would behave with waste water treatment. Thornthwaite, Parmlee, and others associated with the Seabrook project repeatedly warned that, "woods irrigation is a scientific undertaking which requires careful initial study and planning as well as continued reinvestigation if it is to be a successful long-term venture."

As the maxim was used by industrial irrigators, they contributed to the science and art of land treatment. In many instances, their data and experience might serve as basic guides for towns and cities seeking relatively inexpensive, dependable methods for tertiary sewage treatment.

The Campbell Soup Company became a leader in

land treatment, with installations at several large plants around the country. The company solved many of its own waste water problems through research that led to improved techniques for everyone.

Campbell's most talked-about system was built in the early 1960's adjacent to its mammoth Paris, Texas, plant ("22 acres under roof"), which discharged two to three million gallons of waste water per day with a BOD rating three or four times higher than the typical raw domestic sewage of a municipality. The Paris plant was on a six-hundred-acre site where bad soil management during an earlier cotton boom had left the acreage worthless for farming. It was deeply gullied and abandoned to native weeds and other vegetation.

In 1960 Campbell's engineers and Thornthwaite's consultants, including Parmlee, began turning the land into a waste water treatment system. Earthmoving equipment reworked some five hundred acres into gently sloping terraces which were planted to Reed canary grass. They were to serve in an overland flow system with waste water delivered through sprays along the top of the contoured watershed and purified as it trickled down across the terraces through the grass. It was somewhat comparable to the system used for the Borough of Croydon in England sixty or seventy years earlier, and for winter treatment on the Melbourne Board of Works Farm.

By 1967 all five hundred acres were at work purifying all the waste water from the nearby Paris plant. The degree of purification was exceptionally high. The large amount of BOD, for example, was reduced by 99 percent, and phosphorus and nitrogen removal was close to 90 percent. The system worked all winter long, even though the fields turned to sheets of ice. Less waste water was applied at such times, but it was still well purified as it trickled down over the frozen slopes. The effectiveness of the system was confirmed by an exhaustive two-year joint study by the Robert S. Kerr Water Research Center of Ada, Oklahoma, North Texas State University, and C. W. Thornthwaite Associates. Their data were also valuable for land treatment programs in general.

The Reed canary grass planted on the slopes was more than just a treatment fixture. The soup company harvested the hay, which contained up to 23 percent crude protein and twice the mineral content of other local hay. In feeding tests, cattle found the Campbell's crop more tasty than others. Eventually the hay crop brought a small financial return, while its harvesting simultaneously removed nutrients brought to the grassy slopes by the waste water.

After the Paris land treatment was in operation, Campbell's Director of Environmental Engineering, L. C. Gilde, reported that the system had cost about a third the price of an adequate activated sludge plant. Even so, the land treatment method did a much more thorough, dependable purification job — and with hay crops to boot.

On most of its several such land treatment systems the soup company found that savings did not end with the purchase price. Equally or more important, they were easily and inexpensively maintained. Instead of having to deal with complex technology, company engineers and employees were essentially concerned with relatively simple, routine farming problems. There was almost no chance of perplexing break-downs that require highly skilled technicians to diagnose and repair. Above all, the land treatment systems could take large, sudden peak loads of highly polluted waste water without the kind of total collapse that can occur in the concentrated biological processes of conventional treatment plants. Even when the land systems did go wrong, it happened slowly, so that purification didn't abruptly disappear.

Any assessment of such industrial land treatment systems must also consider esthetic values, for a land treatment system may beautify the area chosen for sewage purification. Another of Campbell's land treatment systems is a marvelous example. While it treats the potent wastes of a chicken processing plant at Chestertown, Maryland, it is as much a bird and wildlife sanctuary as a sewage treatment process. It also uses the overland flow technique to treat up to seven hundred thousand gallons of waste water per day on some eighty acres of high grass on gently sloping land. Constant tests show the water receives a remarkable degree of purification before draining into nearby Morgan Creek.

While some of Campbell's land treatment fields are carefully mowed and groomed, the Chestertown system remains relatively wild at the request of the plant manager, Jerry Gardner, who has long been a bird bander associated with the U.S. Fish and Wildlife Service and an avid birdwatcher for his own pleasure. Gardner has turned the eighty-acre treatment area into a strict sanctuary for birds and wild animals. It has attracted far more winged species — including bob-o-link, many of the sparrow family, thrushes, quail, meadow larks, and indigo buntings — than other open land in the area.

"They love the sprinkling water at the top of the area," says Gardner as he conducts a tour of his unusual waste water treatment system. "Birds also come for the grass seed and the berries. They're abundant here." Once in a while on such a tour the walkers flush a deer or rabbit. "We've got a number here," explains Gardner. "We've also got some bobcats, otter, and plenty of woodchucks."

At the bottom of the slope he shows the visitor the test site for the effluent which leaves the sanctuary as clean water that will not pollute Morgan Creek.

If such large, potential polluters can turn their highly potent waste waters into wildlife sanctuaries, consider what many of our municipalities, with less potent sewage, might do with open space and green growth to save us from our present situation of increasing water pollution.

PART 5

Will the Public Accept?
Fact and Folklore

13

The Secret Gold
of Golden Gate

For years Benn Martin, a civil engineer, said very little about his job in beautiful Golden Gate Park in San Francisco because he felt publicity could only harm what he was doing. But then on a social occasion he inadvertently revealed his role in the park to a news reporter, and soon a story about it appeared in a large city daily.

Martin, whose secret really wasn't a secret, was the operator of a sewage treatment plant in Golden Gate Park. The raw sewage came from the city, and the effluent and sludge was used in many of the park's most public areas for water and fertilizer. It had been that way for many years. In fact, the sewage effluent, treated with chlorine to kill disease germs, also helped maintain the water level in several of the park's lovely lakes. As a matter of fact, thousands of visitors, children, starry-eyed swains, and the elderly, had rented boats for carefree rides on the park's Stow Lake, a good share of which was sewage effluent. And at times nearby Huntington Falls, one of the park's prettiest sights, was entirely sewage effluent.

For several years Martin feared that San Franciscans might be outraged if they found out how much of their park's water came from their sewage, even though its use was completely approved and constantly checked by health authorities. The news story proved the assumption false. Nothing bad happened. Actually the report brought praise for making good use of the effluent. Today the practice continues, with many benefits and no serious problems for the park, one of the largest, most beautiful in any city in the world.

The story of Golden Gate Park is one of the more famous to prove that we have a tremendous resource in sewage. But the matter of public acceptance is even more significant, for the sewage water has been used for decades under the noses of millions of people. Many haven't even been aware of it, but a lot have, and they've completely accepted the fact that their park is all the better for turning their wastes to a wealth of green growth.

Whether or not the public elsewhere will approve the reuse of their own sewage remains a key question, especially among public officials. They often assume the answer is "No!" and fail to take the lead necessary to turn our wastes into wealth. Golden Gate Park has been proving for a long time that the answer can be "Yes!"

In 1868, when more than a thousand acres were purchased for the park, the property was considered "the largest, heaviest white elephant San Francisco had ever owned." The land, running back from the Pacific Ocean, was called "the great sand waste." It was mostly windswept dunes with hardly a tree. Native vegetation simply couldn't find a footing in the coarse, shifting sand. When most other cities were carving parks from established woodlands, it was hard to believe that San Francisco's city and county officials would purchase the great sand wastes for $800,000 expecting it to become a park.

Two men made the expenditure worthwhile. The first was an engineer, William Hammond Hall, who worked at the immense conversion job for twenty years. The second, and most famous, was John McLaren, an ornamental horticulturist who served as park superintendent from 1889 to 1943. These two men, with an army of gardeners, using picks, shovels, wheelbarrows, and horse-drawn wagons, capped the shifting sands with a rich humus which still supports hundreds of varieties of trees and shrubs, mostly imported from Australia, New Zealand, and China.

In the early years the main ingredient for the soil-building program was the city's street sweepings, rich in horse manure. The sweepings were transported in work cars on the city trolley lines to a special spur track leading into the park. A little at a time the street wastes helped give the sand dunes a rich dark topsoil.

Horse manure was more easily obtained than water in San Francisco, where a brief winter rainy season is the exception to a long, dry spring, summer, and fall.

The long watering season required more than one hundred thousand gallons of irrigation water per day, which in 1876 cost twenty-three cents per thousand gallons, a tremendous price in those days. Later McLaren drilled his own wells, but he never could afford enough water for the park building job.

When sewer lines were first laid around the park after the great earthquake of 1906, McLaren realized that they brought a magnificent resource; indeed, one line was laid directly across the park. He tapped it and diverted the flow of raw sewage to ditches for irrigation around the middle of the park. But the raw waste water couldn't be managed as it was on the old land treatment farms, and the ditches frequently smelled. People complained, but McLaren refused to give up his new-found irrigation source which also brought valuable organic ingredients for the extensive soil-building program. He kept using sewage, but he limited irrigation to the places and times he could get away with it.

When conventional secondary treatment plants came along, the superintendent realized that here was the chance to make use of the sewage resources much more freely. The problems of health and nuisance odor could be eliminated without forsaking the water or all of the fertilizer value of sewage. A plant capable of treating a million gallons a day was completed in the park in 1932. Henceforth the effluent, chlorinated to remove disease-causing organisms, was used widely and liberally in Golden Gate Park, minus the old complaints.

The park is said to represent "the first planned use

of waste water for recreational purposes to become operational in this country." The secondary effluent was used to fill several ornamental lakes. It was applied to shrubs, flowers, trees, and lawns, including a large polo field. A reservoir built on Strawberry Hill stored the waste water for distribution as required for the wide range of uses.

The effluent, however, was only one beneficial product from the new sewage treatment system. The sludge was dried and used for soil conditioning. The supply was supplemented in 1938 when one of San Francisco's large primary sewage treatment plants was completed in the southwest corner of the park. While its effluent was dumped into the Pacific Ocean, tons and tons of the sludge were used in the park. The Golden Gate gardeners found that it helped soil hold moisture so well that watering was reduced 25 to 50 percent in some cases. At the same time the sludge contributed sewage fertilizing ingredients that stimulated green growth.

The financial advantages were great. In the 1960's the park's treatment plant effluent cost just under ten cents per thousand gallons, compared to over twenty-three cents paid for city water. Even the park's own well water ran nearly twenty-one cents per thousand gallons.

But the low cost of effluent was only one way of looking at the financial advantage. Along with the sewage sludge, it brought so much nitrogen, phosphate, and potash to the park that the gardeners seldom had to order commercial fertilizers—and for a thousand-acre park that was a major saving.

This story of how Golden Gate Park was built on horse and human manure remains incredible to many people. It's a surprise to learn that our own wastes can be tolerated in such an open fashion. But it was only the beginning. By the 1940's the example of Golden Gate Park was followed in other public places, at fancy hotels, on college campuses, and especially on golf courses, including some of the nation's most exclusive links.

Consider, for example, the case of the plush Flamingo Hotel in Las Vegas, Nevada. In 1946 water was badly needed to maintain the beautifully landscaped grounds, and the hotel turned to its own sewage for a supply to irrigate thirty-five well-manicured acres, including the establishment's attractive front entrance. The hotel sewage was treated and stored in a five-thousand-gallon reservoir for use as needed. Sludge from the plant was also utilized by the Flamingo's gardeners. After nearly a decade of effluent irrigation, a report claimed there had been no health or legal problems — and obviously no trouble from the guests.

In four years of study at the University of Florida, students may or may not learn that their campus has been greener and prettier since 1947 because of their own sewage. The grounds at Gainesville received some five million gallons of secondary effluent a month from the university's own treatment plant, and, according to the Grounds Department, the nutrients made their grass, flowers, and shrubs greener than ever.

The same results have been repeated over and over

—with California marines playing baseball every day on a field irrigated every evening with effluent from their own sewage; with municipal golf courses in numerous communities being watered with sewage effluent while the unmindful players concentrate on shaving their scores; and even with the dead of one city being laid to rest under some of America's richest, greenest sod, maintained by the living community's sewage effluent.

All these examples, which date back two, three, or more decades, indicate that people will accept the reuse of their own wastes, yet old assumptions to the contrary remain a strong force working against the municipal use of land treatment. It's an odd corner of public psychology that needs much more airing before the untapped resources of sewage can be widely used.

14

A Break
With Wrong Assumptions

Most people would assume that Americans wouldn't tolerate the Oriental practice of applying "night soil" (human manure) directly to the land. Yet this is what happened for decades from one end of the nation to the other with hardly a dissent. United States night soil was (and still is) distributed from the open-ended toilets of passenger trains carrying millions of people over thousands of miles of tracks. The practice was safe and proven so by authoritative health studies. It was relatively nuisance free. People accepted it.

The subject of what people will or will not accept regarding human waste has many such paradoxes, for much of it is involved with widely held assumptions that are frequently false. Unfortunately the question has seldom been explored by formal research.

132

Yet obtaining public acceptance is, of course, crucial to any attempt to realize the benefits of land treatment of sewage. The proper application of sewage wastes to land can be accepted, as we've learned from numerous case histories. But many public officials, community leaders, and concerned citizens who could be instrumental in developing land treatment systems dare not try for fear that people simply won't condone such practices.

The answer to this dilemma is a matter of building public trust primarily by improving the science and art of land treatment and then educating citizens by word and demonstrations as to the possibilities for effective, safe, nuisance-free sewage disposal through the use of soil and plants, which in turn can offer added public benefits such as open space, green growth, and clean water.

In many ways the history of municipal sanitation and water supply illustrates that when public trust is built on safe, nuisance-free performance, the practices are widely accepted even though they seem to conflict with assumptions to the contrary. This explains the paradoxes.

For example, our attitude toward human waste, as it is widely assumed to be, hardly fits with the fact that drinking water for millions of Americans comes to their taps over a short, direct route from sewage. Many cities and towns draw their domestic water supplies from rivers that are simultaneously used for sewage disposal at upstream communities. As a matter of fact, a single river may serve as both water supply and sewage disposal for several towns in a line. The water users assume that the moving stream's natural cleansing power will make it fit for human consumption, or if that isn't enough, the municipalities install "water treatment" plants to purify the river water on its way into town. People, of course, have long been aware of this relationship between sewage and water supply, as illustrated by a classic sign in public restrooms urging the users to flush because the next town down-river needs the water.

In 1925 G. D. Norcom, writing in the *Journal of the American Water Works Association,* discussed the problems of sewage and water supply in a growing urban society. "Eventually we arrive at a point where it is difficult to say which is sewage and which is water," he stated. "Surely each city lower down on a stream must be drinking a portion of the purified sewage from all the towns higher up."

This was dramatically illustrated at Ottumwa, Iowa, where the water supply came from the Des Moines River, which was heavily loaded with effluent from the upstream city of Des Moines. In the winter of 1939–40 the river was solidly frozen, and the badly polluted water flowed, in effect, through a closed conduit from Des Moines to Ottumwa, so the natural purification of water through contact with air and sunshine was largely inoperative. Meanwhile, the unfortunate Ottumwans learned unforgettably the extent to which their drinking water supply was linked with Des Moines' sewage; it was unfit for consumption until the ice went out.

A close relationship between sewage and drinking water supplies still exists all over America. In 1966 a government study showed that the average river used as a domestic water source contained 2.4 percent waste from upstream communities, and, in some cases, the figure was as high as 18.5 percent.

Still, over the years millions of citizens have rightfully trusted municipal officials to provide safe, pure drinking water. Even when the water could be associated with sewage, it didn't matter much as long as tap water continued to taste, smell, and look relatively good, or as long as someone — perhaps a journalist — didn't stir up fears, with or without validity. Indeed, under certain circumstances, citizens authoritatively assured of their water being safe have often shown amazing tolerance to the fact that it is being reclaimed directly and immediately from sewage.

The ultimate example of this came from Chanute, Kansas, in 1956, near the end of the state's worst drought. Chanute, which is located on the Neosho River, was the eighth town in line to use the river for drinking water and sewage disposal. But that summer the flow stopped, and Chanute was left with only a pool of water caught behind a local dam. Town officials were determined not to lose the precious supply, so they arranged to use it over and over again. As usual, the water was purified in the water treatment plant, distributed to the people, collected and treated as sewage, and finally returned to the pool behind the dam, ready for another such circuit through the community.

This recycling continued for five months with the officials making a heroic effort to keep the domestic water supply up to state health requirements — which they did with the state reluctantly approving. Chanute citizens still kept their trust in the town officials, and used the water, until the increasingly bad taste and appearance were simply too much to take. When the drought ended, people were overjoyed to return to their normal river water, even though it had seven known uses before arriving at Chanute.

A carefully designed system of recycling sewage water has been well accepted by the citizens of Los Angeles County since 1962, when the Whittier Narrows Water Reclamation Plant was constructed. Some twelve million gallons of secondary effluent a day are mixed with natural runoff and with Colorado River water (both unpotable), and purified by filtering down through the earth from surface spreading basins located along two river beds. Below ground the purified effluent becomes part of the groundwater supply used for drinking water by a large area of Los Angeles County. In fact, the Los Angeles County sanitation districts actually sell the infiltrated effluent to the Central and West Basin Water Replenishment Districts, the organization that maintains wells and pumps to supply domestic water to the area.

In its first ten years this "groundwater recharge" project reclaimed nearly 500 billion gallons of water from sewage effluent and sold it for $2,250,000, which helped pay for the Whittier Narrows treatment plant and its operation. Extensive scientific research on the project showed that purification was effective and safe, and, as one of the investigators concluded, "The passage of water through soil is one of the most effective and economical purifying mechanisms known to man." It allowed a lot of Californians to enjoy a great deal of water which otherwise would have been abandoned had county officials believed the assumption that people won't accept the reuse of their own sewage.

Further proof that the public will accept the reuse of sewage is evident from at least two remarkably successful fertilizer products that are actually sewage sludge from municipal treatment plants. One is sold only on the West Coast; the other is marketed all over the United States, Canada, and in several other countries.

The West Coast product, Nitrohumus, has been sold since the 1920's when a surveyor, H. Clay Kellogg, discovered that luxuriant growth along the Santa Ana River was stimulated by sewage sludge from a nearby treatment plant. He carried some home, tried it on his lawn with fabulous results, and decided to go into business selling sewage sludge. He obtained raw material from the Los Angeles sludge-drying beds and sold it in bulk to citrus fruit growers. At first the farmers were wary about the human waste product being fit for fertilizer, but the excellent results were convincing, and Kellogg soon had a thriving business. After World War II the Kellogg Company marketed sludge in bags for home use with the trade name "Nitrohumus." The firm still sells tons and tons of Los Angeles sludge. It's the fertilizer most widely used by homes, nurseries, and landscape contractors in southern California.

The nationally sold product is "Milorganite," and it accounts for practically every pound of sewage sludge produced by the Milwaukee (Wisconsin) Sewerage Commission, which treats nearly a half billion gallons of sewage a day in two huge plants. The dried sludge may add up to eighty-five thousand tons a year, and Ray D. Leary, the commission's chief engineer, says, "If we had more, we could use it."

Milorganite's origins go back to 1920 when a young agricultural student, O. J. Noer, received a four-year scholarship, sponsored by the city of Milwaukee, to the University of Wisconsin to study the fertilizing ingredients of sewage sludge. The research, which confirmed the values we've already discussed, led Noer to organize the sale of dried sludge by the Milwaukee commission. A $100-prize contest produced the product name, and the sale of Milorganite increased and spread. Automated equipment now loads about a dozen and a half fifty-pound bags a minute. They are shipped to sixty dealers across the nation who are held to quotas because the heavy demand outruns the Milwaukee production.

Does anyone ever object to the real origins of Milorganite? "Not at all," says Leary. "We get thousands of letters from gardeners praising the stuff. It doesn't burn their plants. It's organic, and that's important."

Many of the common assumptions about this question of public acceptance were tested in an extensive study in the late 1960's by James J. Johnson, then with the University of Chicago Department of Geography. He explored how people would react if they learned that their communities were going to use reclaimed sewage. Would they let their families drink water they knew had been renovated from waste water? How would they feel about the cost of the re-claimed sewage water? He conducted the study through extensive, carefully designed interviews with citizens of a broad economic and social range in Camden, New Jersey, Philadelphia, Cincinnati, Tucson, and Portland, Oregon.

Johnson's detailed results revealed that the majority of people would accept the use of renovated sewage water if the renovation were properly handled. Where future water shortages faced their areas, the interviewees appeared more willing to reuse water. Also, as proved true in Chanute, people whose existing water sources were of poor quality showed more willingness to reuse their sewage water.

But Johnson reported that education was the most significant factor in acceptance or rejection of waste water reuse. The more educated the individual the more likely he was to be willing to use and drink renovated water. In this same respect, the study revealed that the better acquainted a person was with the way in which water was reclaimed the more likely he was to accept it for reuse.

The University of Chicago investigator concluded that public education could be the essential key to our taking advantage of the resources in sewage that are now lost by present disposal policies. Johnson felt his study "may well signal the importance of public information programs as guides in seeking more favorable public response."

Perhaps most significant was an indication from the research that key public officials need education on public attitudes most of all. "It would appear," Johnson wrote, "that water managers know very little of consumer responses concerning renovated waste water, but generally consider that the public would not accept it."

Of ten managers interviewed in Philadelphia, all thought the public would disapprove of reusing their own sewage water. Similar responses came from officials in other cities. Their assumptions, which are contrary to what Johnson's study indicated about the true public attitude, would of course greatly reduce the chances for sewage ever being reused in their communities.

As we have seen, there are public officials who have refused to be blocked by false assumptions about public acceptance, and they've been able to turn their waste waters to forms of wealth for everyone's benefit. One of the best examples of all is America's fastest growing city, Colorado Springs, at the foot of Pike's Peak, where community officials have, with the citizens' understanding and approval — indeed, praise — put sewage effluent to work all over town.

The city, which has constantly faced a water shortage, has two water systems. One carries the community's potable water, which costs about fifty cents per thousand gallons. The other transports well-chlorinated sewage effluent, which sells for about seven cents a thousand. This secondary water system was started in 1964 to help the local highway department beautify a new expressway by spray-irrigating grass and other plants along the borders. This demonstration of the possibilities, plus constant assurance from sanitation officials of the safety of the practice, led more and more institutions to buy effluent, and the demand eventually warranted the secondary piping system.

By the 1970's effluent was irrigating and fertilizing grass and plants all over the city, and many of the buyers were making up for the expense of the waste water by savings in commercial fertilizer. The broad, green lawns of a large trade union retirement home were irrigated and fertilized by waste water. Because of the enriched irrigation supply, there were now greener city parks and several golf courses, including the exclusive Kissing Camel course. A large cemetery was irrigated with sewage effluent. And the entire campus of Colorado College was turned from a fading green that often went brown to a continual bright green.

Even the very heart of Colorado Springs was beautified by effluent irrigation. Three of the busiest, widest streets extending some three miles down a gentle slope through residential and business areas have center strips of grass several feet wide and interrupted only by the break at cross streets. Each of these block-long segments is flood-irrigated with secondary effluent, which is released at the high end and allowed to flow through the grass down the incline to where it is collected at the next cross street.

Has anyone ever complained? Jim Philips, Superintendent of the Sewer Division, recalls only one woman who said she objected to the delivery of sewage water to her neighborhood. Her argument had little substance, and though it stirred up some public fears, Philips was able to allay them with evidence that the wide public use of effluent was not a nuisance or a hazard to public health. He also had data from test wells to show that effluent cleaned by plants and soil as it filtered into the earth was not contaminating groundwater supplies. Nothing ever came of the complaint. "Of course, one loud, emotional voice gets a lot of attention," Mr. Philips said, "when actually such outbursts are far outnumbered by the words of praise we receive for our system."

Philips believes the key to the intelligent reuse of our waste water is public education. In addition to educating the adult population, he has been working with the city schools, and now sewage treatment and reuse is part of the curriculum, which includes two field trips per student around the community sewage treatment system, once in the sixth grade, once in the eleventh.

Indeed, nearly everyone who seriously proposes the reuse of our sewage resources eventually comes around to recognizing that technical problems are not the most formidable obstacles in the way of advancement. The main problem is to break away from the old assumptions which block us from turning to new, beneficial alternatives. It can be solved, as we have seen—and as we're about to discover, in the most remarkable project of all.

15

How They Saved
the Sewage of Santee

A unique festival is held each June in Santee, California. It features a long, colorful Saturday morning parade that starts from a shopping center and ends at the town's recreational lakes. For the remainder of the weekend, as many as ten thousand people celebrate the construction of eight beautiful man-made lakes laid out in a line for the better part of a mile. At this "Festival of the Lakes," people fish, picnic, boat, and swim, with little concern that nearly every drop of the water is supplied by Santee's municipal sewage system.

In the 1960's Santee, with fourteen thousand inhabitants, astonished many of the nation's authorities on water and sewage who had firmly convinced themselves that the public would never agree to the reuse

143

of their own sewage, especially for swimming. Santee's citizens not only accepted the idea but became irate when it was not promptly put into effect.

The technical means of purifying their waste depended in the final analysis on the marvelous filtration qualities of soil, but the real key to success of the system was a program of public education led by the former proprietor of an appliance store who had come to the water and sewage field without the fetters of false assumptions about people and sewage.

The unusual story began in the late 1950's when Santee was asked by nearby San Diego to join a proposed pipeline project for sewage disposal into the Pacific Ocean. At that time, the town could sign a forty-year contract at an annual rate lower than would be possible once the project was completed. Other communities took the offer, but Santee's county water district had a young manager, Ray Stoyer (the former appliance dealer), and a board of directors who were entertaining some other ideas for the town's daily waste water output of nearly a million gallons.

In Santee, which is located in the arid San Diego River basin, water was a precious commodity piped some 300 miles from the Colorado River at a cost of $17 per acre foot. Furthermore, the cost was increasing. California water authorities estimated it would be $75 to $100 per acre foot by the 1970's. With such a valuable resource, Santee's officials thought that perhaps they should consider how to keep their water rather than how to pay San Diego to throw it away.

Soon Stoyer and his board were learning some of the elementary facts about raw sewage that we've already reviewed: for instance, that over 99 percent is pure water and that the last fraction of a percent can be removed with the help of nature by lagoons and the power of soil as a complete purifier. So, unhampered by all the problems then being encountered in the search for tertiary treatment, the Santee officials decided to capture the fresh water in their sewage and put it to good use. They already had a new, fairly good primary-secondary treatment plant. Nature, they figured, would help do the final cleaning job of the secondary effluent. The water in sewage was much too valuable to throw away. "We decided," said Stoyer, "that long before a forty-year disposal contract with San Diego ended, waste water could become a water resource, not a liability that you pay to get rid of."

In fact, Stoyer and his colleagues decided that as much as anything their town, which has less than ten inches of rain per year, needed some freshwater lakes. People were driving hundreds of miles to go fishing in fresh water. The reclaimed sewage could be used to green up an area around the lakes for picnics and playgrounds. The lakes would be great for boating, which people now couldn't enjoy except in a limited way at the ocean near San Diego. And why not figure on using the water for swimming? Everyone agreed that the recreation idea was great, and Stoyer set his mind on perfecting it.

But the stumbling block was less technical than psychological. There was little precedent for a town or city making outright public use of its waste waters for recreation, especially for swimming. Offering the public the use of reclaimed water was something officials had always avoided, even if they had thought it possible. One illness traced to the water would stir up immeasurable public wrath.

Stoyer, however, argued hard in favor of a reclamation project. He asked if the idea was really so unusual. A lot of municipal water supplies, he said, contain what amounts to reclaimed sewage water, although it is not advertised as such. He brought out that many towns and cities draw water from a river into which upstream communities dump their waste. He talked about the idea with everyone. The key to it all, he concluded, was public education. Once people were shown that reclaimed water could be completely safe, they would accept it for many purposes.

Finally, convinced that sewage reclamation should be their goal, the water district's board voted not to join the San Diego system. Then the new, untried idea of a town reusing its waste water had to succeed or there would be no place to dump the million-gallon daily output of sewage effluent from their new $500,000 treatment plant. The effluent at the time was pouring into nearby Sycamore Creek, but that would soon have to stop, for state antipollution regulations were becoming more stringent. Santee officials had to get moving, so they decided to build a lake or two with the effluent and let people see what might be possible if the recreation idea worked with their help.

"We decided to associate what we were doing with pleasant things," explained Stoyer. "If we could tie in our activities with such happy things as swimming, fishing, picnicking, pretty green lawns, trees, and sailboats, it could have a lot to do with public acceptance.

The creation of a lake, even though it wasn't approved for use, would provide the opportunity for great numbers of people to come down and look at the water, smell it, and examine it for themselves. They'd see it was okay. Nothing will convince an individual more quickly than seeing for himself."

Stoyer visited a gravel mining operation which was bulldozing pits along the Sycamore Creek valley, on the western edge of the growing community, next to some of the town's newest residential developments. The area was one of the worst sights in town. The man who owned it agreed that the gravel mining operation could produce lake sites as well as gravel pits. The pits would make a good place for the reclamation project because they were right next to the sewage treatment plant, and the gravel man could build a golf course on some more land of his below the proposed lakes, using the spillover to provide low-cost irrigation water.

The miner agreed to turn over his depleted pits to the town, and soon Stoyer's men and equipment shaped the devastated area into three basins. The first was filled with effluent to serve as a lagoon for further water purification. After it was filled up and had been exposed to air and sunshine for a time, the water—which was now remarkably clean, but not by any means clean enough for swimming—was pumped into the other basins.

At this point Stoyer used the old and simple carrot-on-a-stick psychology with his fellow citizens. He planted grass around one of the new bodies of water which was then used also for spray irrigation—and

which already had trees, spared by the gravel miner —
and soon the area looked sensational against the
brown, parched countryside. Stoyer added some picnic tables and put a boat or two on the water. Then he
imported some swans to join the other birds already
attracted by the water. Altogether it was a marvelous
sight. But then the district manager surrounded the
delightful place with a high wire fence — and invited
everyone in town to come for a look, and a good smell,
if they feared they might smell something bad. But
they couldn't go inside the enclosure.

All this time Stoyer was promoting the recreation
idea in all possible ways. He used the newspapers for
every kind of story that came to mind. He spoke to all
the civic groups in Santee and in every other place
that would invite him. He figured distance made little
difference; word would get back to town somehow.
For a while he was number-one public speaker on the
local lecture circuit from San Diego east for many
miles. "In those days," he said, "I'd even see three or
four people on a porch, and if they'd let me, I'd give my
speech on the reclamation project."

The strategy worked. People started to get impa-
tient about the fence keeping them from the new lake.
One day Stoyer had a telephone call from a frustrated
viewer. "Say, Mr. Stoyer, why *can't* we go in there
and enjoy the lake?" said the voice on the line. "Too
bad for that place to be going to waste when people
could be using it."

Stoyer was delighted by the complaint. "Well, it's
our aim to use it," he told the caller, "but we have to
make absolutely certain that the water is safe. The

The height of Penn State corn, spray-irrigated with secondary sewage effluent, tells the story. (*Penn State*)

In the wintertime spraying of sewage effluent moves to Penn State woodlands where the soil can continue to accept water for purification. (*Penn State*)

(Above) One of the Santee lakes, where every drop of water was only recently sewage from nearby Santee, California. The pool (below) is filled with water from the lakes. Santee's residents have completely accepted the source of the water. *(Leonard Stevens)*

A Campbell Soup Company spray-irrigation field treating potent waste water from the nearby food processing plant. (*Campbell Soup Co.*)

One of several percolation beds at Santee where secondary sewage effluent is filtered through the earth for complete purification. (*Leonard Stevens*)

This terraced land treatment system, on 500 acres at Paris, Texas, treats highly polluted waste from the huge Campbell Soup plant. The system uses the overland flow technique. (*Campbell Soup Co.*)

Spray-irrigation rigs, which pull themselves slowly along cables that run between the wheels, apply waste water to a newly planted field of eucalyptus at Walt Disney World's Living Tree Farm in Florida. The "cannons," which spray the water, deliver about one million gallons per day to the land. (*Walt Disney Productions*)

A portion of the tertiary treatment plant at South Lake Tahoe, California. The tower in the background removes nitrogen. (*Leonard Stevens*)

Three bottles of waste water at South Lake Tahoe: raw sewage (left), secondary effluent (center), and tertiary effluent (right). (*Leonard Stevens*)

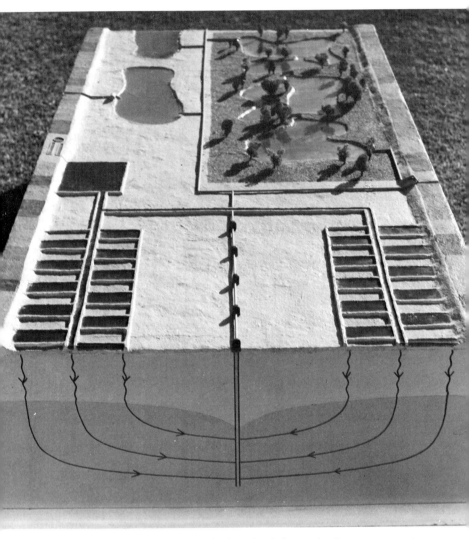

A model of a proposed Phoenix, Arizona, land treatment system. Secondary effluent from the city will be spread on filtration beds (left and right foreground) for purification as it seeps down and to the center of the dry river bed. Pumps (center of model) will then raise the purified water for use in the recreational area (right rear). (*Rio Salado Project*)

An aerial view of the Flushing Meadows Project, Phoenix, Arizona. In a desert area outside the city, secondary effluent is applied to grassy beds for infiltration of the earth. Nearby test wells allow scientists to analyze the purity of the water. (*U.S. Agricultural Research Service*)

A Pasveer oxidation ditch (racetrack) at Stayton, Oregon, provides secondary treatment as two horizontal rotors (one near the footbridge) force waste water around the ditch. A settling tank (circle in foreground) collects sludge from the effluent. (*Lakeside Equipment Corp.*)

Two new treatment systems of Dutch origin, back to back, called Carrousels. The system is evolved from the Pasveer oxidation ditch, but in this case the flow of water through the unit is guided by baffles. Vertical rotors near the center move the waste water. (*Envirobic Systems, Inc.*)

Two newly filled biological treatment lagoons at Muskegon, Michigan, shortly after the land treatment system began operation. Twelve floating aerators at the center of circular areas of foam introduce oxygen to the eight-acre ponds. Mixers at the center of the dark patches keep the solids suspended for later removal as sludge. These lagoons, with one other not in view, provide secondary treatment to raw sewage. At the left, partially in view, is a huge storage lagoon, from which effluent will be drawn for spray irrigation of nearby agricultural land. (*Teledyne Continental Motors*)

A spray-irrigation rig at Muskegon applies waste water to a field of corn. The rig is moved on wheels around a central point, making the radius of a huge circle. Waste water is sprayed downward to prevent tiny droplets from being carried off by the wind. (*Teledyne Continental Motors*)

state has to approve before we can let people in there. We're working on getting the approval. Just be patient. It has to be done right."

More such complaints made Stoyer feel his methods were working. "All the way along," said Stoyer later, "we got people pushing us to let them use the lakes, rather than our trying to persuade them to come in. From the time we showed them the first lake, the public was always seeking to use the lakes in different ways, and we were always saying, 'Well, no, we are not going to let you use it until the studies that are going on have proven it is safe for you.' This not only built public desire and acceptance for the project, but also put people at ease about using water that was recently a part of the town's sewage."

Meanwhile Stoyer kept working at the technical side of the project. His own tests of bacteria and sewage nutrients showed the cleaning job had to continue before the state of California would ever approve the new lakes for recreation. To solve the problem Stoyer went back to nature, this time to the soil.

From the lagoon, he pumped the water about a mile up the gently sloping Sycamore Creek valley to six shallow, half-acre gravel beds, "percolation beds," cut into the valley's eastern slope. The water percolated through the gravel and through the earth a few yards down the slope to a collection ditch paralleling the beds. At this point Stoyer's tests showed hardly any bacteria and only a small harmless fraction of the sewage nutrients. From the ditch the water was sent through a small chlorination plant as a super-precaution against the remaining bacteria. From there the water was dumped back into the uppermost of the lakes, now four in number counting the lagoon. They formed a line of lakes up the valley with the lagoon as first in line. Starting with the uppermost lake, the water spilled from one lake into the next down the valley until it reached the second in line, the one above the lagoon. From this lake it was released temporarily to the creek.

Santee still had to jump some formidable hurdles before the San Diego County Department of Public Health, headed by a strict physician, Dr. J. B. Askew, would permit extensive recreational use of the proposed lakes—especially for swimming. The project would require some basic scientific research to develop strictly foolproof methods for testing the purity of the water. The big problem was one of identifying viral organisms that could endanger health. Available viral tests at that time were not good enough for the Santee project, but for the town to tackle the research to make the tests was financially unfeasible. On the other hand, if the research could be done, the results would be valuable to hundreds of towns and cities—all of which indicated that higher governmental agencies might be interested in helping Santee. Stoyer decided to look for such help.

He found plenty of interest and possible assistance from county and state agencies, but everyone realized that there were major problems, especially money, which really called for help from Washington. When Stoyer applied for federal assistance, it was approved and supposed to come, but didn't. Too much red tape stretched between Washington and Santee. One day a disturbed Stoyer penned a plea for help directly to President John F. Kennedy. By return mail a White House assistant expressed the president's interest in the pioneering project and said Mr. Kennedy would give it attention. Assistance soon came in quantity and quality. Somehow Public Health Service officials had suddenly been made to realize that Santee was a lone town with vital leadership willing to try what others had feared to attempt.

By April 1962, a memorandum of understanding was signed by the U.S. Public Health Service, the California Department of Public Health, the Santee County Water District, and three other state and county agencies, all of which agreed to work together on the Santee water reclamation project. With the agreement came money, equipment, and personnel from all the agencies. A mobile laboratory and a number of technical people were provided by county and state health departments.

Directing the combined effort was a U.S. Public Health Service water supply and pollution control expert, John C. Merrell, Jr. His work was to bring him the highest honor given civil service employees, the Distinguished Service Award.

The most difficult problem faced by Merrell and his people was virus detection. While virologists could locate certain viral organisms in a sample of water, they could not be certain they had identified all that were present. Improved identification techniques requiring a major research effort had to be devised before the lakes could be declared entirely safe.

A microbiologist, Mrs. Beatrice England, working under Dr. Askew's San Diego department, was assigned the virus research. Her months of work, which became an important general contribution to virology, developed methods for isolating and identifying all the viruses present in a water sample. Mrs. England's tests eventually provided the most convincing evidence that Santee's lakes were safe for human contact.

By coincidence the virological tests were conducted as the citizens of Santee were taking the Sabin polio vaccine as part of the famous nationwide mass oral feeding program in 1962–63. Thus it was certain that the human feces in the town's sewage were introducing unusually large numbers of polio viruses to the purification system. While Mrs. England's tests could find them in samples of raw sewage, there was no evidence that a single one of these billions of viral organisms made it past the percolation beds. In fact, most were gone from the waste water before it left the lagoon prior to going to the beds.

Subsequently, Dr. Askew worked out the most stringent test of all for the soil filtration process. He had three gallons of diluted Sabin vaccine injected into the pipe leading to the filtration areas; the effluent passing through the pipe was therefore heavily contaminated with polio viruses. After the effluent seeped down through the soil, it was collected in special wells for virological tests by Mrs. England. She was unable to find that a single viral organism had made it through the earth filters.

Once the scientific data had proved the water's purity, the lakes were opened for various public uses, a step at a time. In April 1963, boating was permitted. Birdwatchers discovered that the lakes were attracting a wide variety of birds. And visitors, many from San Diego and other communities, were allowed to picnic on the grass and under the trees. On the east side of the lakes, real estate values rapidly increased.

During this time, the question of whether the public would accept the waters switched to the question of whether the waters could accept the public. People were anxious to use the lakes for swimming and fishing. The final answers had to come from Dr. Askew, who could not be rushed by public pressure when his county health department had to approve a new, more risky recreational use of the water. He even delayed a planned dedication of the lakes and was scorched by a critical editorial in the local newspapers. But such controversy was good; when the doctor finally said yes, people knew his decision could be thoroughly trusted.

In June 1964, fishing was allowed and anglers took some beautiful catches home to their frypans. In 1965 a recommendation was made that the water, channeled into a separate sandy-bottomed pool, be approved for swimming. Dr. Askew accepted the recommendation, and the swimming program began. At first, only twenty-five swimmers were allowed in the pool at a time and they were carefully checked by epidemiologists for any possible ill effects. The results were favorable and the facility was opened to its capacity. Soon, the 65-by-87-foot pool was not big enough, and Santee had to build a second and larger swimming installation.

The lakes became famous, and water and sewage authorities came from all over the world to learn how this small community had in one stroke solved its problems of water pollution and water shortages. Stoyer, in turn, was invited to conferences all over the country to relate how he had accomplished the unthinkable: getting people to swim in their own sewage water. He explained that the technical part, of course, was important, but the nub of the problem was getting people to think of the pleasant possibilities in the reuse of their own waste water. That naturally took time and skill in human relations, but once accomplished, people not only wanted to take advantage of the otherwise wasted resource — they demanded it.

PART 6

Understanding, Testing,
and Trying
Nature's Purification
Processes

16

The Science of Soil, Plants, and Sewage

Two scientific tests, each conducted for different purposes not long ago, revealed the power of soil and plants for removing the valuable nutrients of sewage.

One was made for the Society of the Colonial Dames of America to help them faithfully restore a famous colonial home in Wethersfield, Connecticut. The restorers knew the house once had an outdoor privy, but they didn't know exactly where it had been located. A soil scientist from the Connecticut Agricultural Experiment Station offered to find out. He tested soil samples from different spots around the yard until he found one area especially rich in the primary nutrients of human waste, especially phosphorus and potassium. There, he decided, was the former site of the privy. The enriched soil had held onto the nutrients for some two centuries.

In the second test Harry J. Eby, a research engineer at the U.S. Agricultural Research Service at Beltsville, Maryland, designed a remarkably simple tertiary treatment plant that used only living grass to remove the waste water nutrients that cause many of our water pollution problems. The system depended upon hydroponics, the science of growing plants in liquid solutions instead of soil. The grass was planted in shallow, rectangular tanks containing a layer of crushed rock. Waste water effluent from a lagoon was allowed to flow very slowly through the stones from one end of the tanks to the other. The grass roots extended down among the crevices between the rocks, giving the growth physical stability. Food for the plants came solely from the effluent flowing through the crevices. With this simple system, Eby was able to remove 70 percent of the nitrogen and 80 percent of the phosphorus that remained in the lagoon-treated waste water. The nutrients were taken up by the grass, and from there, of course, they could have gone back into the nutrient cycle in a meal for a cow.

As the history of sewage treatment reveals, people have known for a long time that soil and plants can clean waste water with great efficiency, but they have not understood exactly how these natural systems work. This limitation made land treatment more of an art than a science until recent times, and naturally it was responsible to a degree for the failures that often were cited by those opposing wider use of land treatment.

But in the past twenty years scientists and engineers have been giving the subject more intense study. They have compiled data that will help municipalities and industries design viable land treatment systems with a wide variety of soil and climatic conditions, and with various volumes and kinds of waste water. As the research proceeds, the results should make it less difficult for a community or industry to custom-design effective land treatment systems with native soils and plants. In the meantime, the growing body of knowledge is already at work in a number of pioneer systems designed to take maximum advantage of nature's marvelous purification processes.

The needed research is not so much a matter of developing new knowledge as it is of applying existing knowledge. A great deal of the data needed to design effective, beneficial land treatment systems has already been compiled by crop and soil scientists. Their findings are now being tried and tested in the waste water field and reported upon through scientific and technical journals, books, seminars, and other avenues of communication.

It is already well established that soil works as a filter of sewage in three ways: physically, chemically, and biologically.

The fact that soil can literally act as a physical strainer for waste water is the best known of its capabilities. However, good and bad experiences have shown that the filtration qualities vary widely, which must be recognized when locating and designing an effective land treatment system. First, the designer must find sites where the earth's filtration capabilities will work for an indefinite time with the volume and kind of waste water to be treated. Then he must prescribe management practices that respect the limits of the soil and preserve its filtering qualities.

The designer must also take into account several chemical processes at work in the soil when it is used for land treatment. The most important ones include ion exchange, adsorption and precipitation, and chemical alteration. Through these processes the chemical ingredients of waste water may be taken up by the soil, released to the atmosphere, or altered chemically. An understanding of how they work and of their limits is essential to effective design and management of a land treatment system.

And the designer must consider the biological filtration processes of soil which are found mostly in its top five or six inches. In this thin layer of earth there are huge populations of microbes, and the numbers increase when organic waste materials are applied. They have the key job of decomposing biodegradable organic materials, thus adding soil humus which in turn aids the allied physical and chemical processes. They can also degrade certain toxic materials or unwanted organic compounds contributed to sewage by industrial sources, detergents, or pesticides. And the microbes have a central role in the important process of nitrogen removal. While there is already a lot of applicable knowledge about soil as a biological filter, more needs to be learned, especially about nitrogen removal. Such information is essential to take full advantage of land treatment under all the different conditions experienced in a wide variety of communities.

The power of these three filtration processes in soil

alone is proven in practice at such projects as Santee and Whittier Narrows, where sewage effluent is reclaimed for swimming and drinking water purposes after percolation down through the earth. The remarkable filtering capabilities of common earth have also been demonstrated in the laboratory. For example, the Connecticut Agricultural Experiment Station at New Haven conducted a two-year experiment that revealed the purifying power of various native soils.

Six were selected with different chemical and physical characteristics to provide a representative sample of thousands of Connecticut acres. Cores of the various soils, each twelve inches in diameter and three feet deep, were carefully lifted from the ground at selected sites around the state. They were encased in plastic and trucked to the state soils laboratory at New Haven where they were mounted upright in steel racks. For two years a synthetic sewage effluent was poured on the exposed tops of the cores and allowed to seep down through the encased soil at a rate equivalent to dousing an acre of land with three million gallons of effluent a year. Drain taps inserted at intervals down the vertical sides of the plastic containers allowed the investigators to sample the water after it had filtered through one or more layers of soil.

All the columns were effective at removing phosphorus, potassium, calcium, and magnesium from the passing effluent, but the soil texture made a difference as to the degree of effectiveness. Coarse textured soils, for example, did not reduce the potassium and magnesium to acceptable levels but were good with phosphorus and calcium. Without the benefits of living root systems none of the soil samples could be depended upon to remove nitrogen. It was converted to nitrate in the earth, and this "nitrogen-nitrate" leached down through the soil cores.

The Connecticut scientists felt the study demonstrated that the state's soil (with crops to improve nitrogen removal) could be used as an effective tertiary treatment system for sewage effluent—providing that the "vast differences in permeability and capacity to remove and store nutrients" were considered in determining the volume of effluent a particular site could be expected to treat.

While soil alone can provide a high degree of purification, its capability is increased by the addition of plants, which, in effect, add another dimension to the filtering ability of land. The combination has become widely known as a "living filter," a term originating with an extensive Pennsylvania State University project that is described in a later chapter. Of course, the effectiveness of the living filter may be enhanced by harvesting and removing the plants from the treatment site in order to take away the nutrients they absorb. But harvesting isn't always essential because some soils and the living organisms they contain are capable of absorbing or removing great quantities of nutrients. Certain prairie soils, for example, can hold several tons of nitrogen per acre. Soil bacteria may also play an important role in nitrogen removal through the process of denitrification.

Plant growth in land treatment is most closely related to nitrogen removal because, of all the primary nutrients of waste water, nitrogen is the one considered most important to growth. Conversely, this means that various kinds of vegetation irrigated with sewage effluent can remove large amounts of nitrogen. At the same time, vegetation will also take up smaller quantities of phosphorus and potassium. But phosphorus, which is considered second in importance to nitrogen for plant growth, is taken up at only a fifth or less the rate of nitrogen.

Corn, which can have a considerable commercial value of its own, can also have a value in nutrient removal, especially nitrogen. An acre of corn may take up around 160 pounds of nitrogen, some 25 pounds of phosphorus and 50 pounds of potassium. Other crops noted for good nitrogen removal include soybeans, alfalfa, and red clover. Reed canary grass, a common perennial grass often grown in hayfields and pastures, is frequently used in land treatment systems because of its nitrogen removal capabilities. It may take up as much as 300 to 400 pounds per acre.

Of course, the filtration values of such crops are limited to growing seasons that may be relatively short in colder climates. As we learned earlier, the porous floor of a forest may provide for cold-weather soil filtration after the growing season is over. In this case the plants of the living filter are mainly trees. But their capability for taking up nutrients is not as well understood as agricultural crops. Nutrient removal by harvesting the crop is definitely complicated so the capability of forest soil to store and remove nutrients by bacterial action is important.

The problem of removal by harvest is revealed by a look at the distribution of nutrients in a single tree.

The highest percentage is concentrated in the foliage, and from there the amounts per parts of the tree decrease in this order: bark, branches, trunk, and roots. Unfortunately the trunk is preferred for commercial use of a tree. The foliage has little value and normally falls to the ground, returning large concentrations of the nutrients to the earth. Only an unlikely, uneconomical army of leaf rakers could remove nutrients by way of harvesting.

However, trees and the unique floor of the forest can still play an important role as a living filter. Despite the concentration of nutrients in the foliage, all the other tree parts collect and hold appreciable amounts of nutrients. Also the unique woodland soil and its living organisms provide an efficient filter during the seasons when other living filters are inoperative.

In general, a well-managed combination of agricultural and woodland irrigation seems to be one of the answers for land treatment systems in colder, wetter climates. In the usual growing season the irrigators turn to open land, with crops carefully chosen and managed for the maximum removal of nutrients. At other times they turn their effluent sprays to nearby forest lands, and within the wooded area they alternate between spraying a tract and allowing it to rest, according to a timetable designed to attain the best nutrient removal.

As already repeated through numerous examples in previous chapters, the idea of the living filter offers much more than waste water treatment. It can provide crops with yields far greater than are expected from growth without the nutrient-rich irrigation. This added value also applies to certain species of trees. Their annual growth when treated with sewage effluent may increase dramatically — which could make the process economically attractive to tree farmers. And finally the living filter is esthetically superior to any technological filter, especially in urbanized areas where the preservation of open space graced with green fields and woodlands is badly needed.

17

The Healthiest Way:
Nature's Way

Anyone who advocates the return of municipal sewage effluent and sludge to the soil for land treatment risks being branded a menace to public health. It's a serious deterrent to public officials and concerned citizens who recognize land treatment as a viable alternative to the troubled technology so widely promoted for water pollution control. The person willing to accept the risk and pursue the alternative needs to know about sewage and health. He should be concerned with two questions:

• Granted that sewage, both raw and treated, can be a health hazard, how serious is it?

• What can nature's purification processes do to reduce or eliminate any existing threat to health?

The answer to the first question is not fully known,

166

although some time-worn assumptions indicate that sewage and disease are practically synonymous.

The relationship of sewage to health is a serious consideration because certain disease germs are waterborne. Historically, the spread of great epidemics, such as typhoid fever, cholera, and diarrhea, was correctly blamed on the pollution of drinking water by human wastes. A classic example was the Asiatic cholera epidemics of 1892–93 in Hamburg, Germany, where 8,605 people died. The extent of the epidemic resulted from the mixing of drinking water and sewage by tidal forces in the Elbe, which served the city as both water supply and a receiving stream for raw waste water. Such disasters led to widespread efforts to keep sewage from contaminating drinking water. The result was one of the great public health achievements of all time.

But meanwhile, sewage and disease became practically synonymous in many minds. It still remains so with a large segment of the public — and with many public officials. However, experience and research have revealed that waste water from modern, healthy communities is not necessarily as hazardous to health as is commonly assumed.

If sewage and disease were synonymous, one would assume that sewage workers would be an unhealthy lot. The opposite seems to be the case, as we learned earlier. Today's operators of sewage plants consider themselves exceptionally healthy people, and absentee records frequently support their contention. For instance, a study in New York City a number of years ago revealed that sewage workers had the lowest rate of absenteeism of a number of occupational groups investigated.

If sewage and disease were synonymous one would assume that animals drinking sewage would suffer dire consequences. The assumption was tested by the U.S. Department of Agriculture at Beltsville, Maryland, in the late 1930's when stock owners initiated frequent lawsuits to keep municipalities from contaminating agricultural drinking supplies with sewage effluent. For six months Beltsville scientists fed a number of healthy swine and cows on diets heavily loaded with either raw sewage or treated effluent and occasional portions of sewage plant sludge. The animals were subsequently observed and tested for another six months before being slaughtered, whereupon the meat was examined. All the tests failed to turn up the least sign of an animal acquiring a disease from the diet—which, incidentally, the swine and cows relished.

While such empirical evidence says that sewage doesn't necessarily equate with disease, there's little scientific evidence defining the true relationship. For that matter some of our strictest health laws and regulations rest on evidence that seems to predate modern ideas about disease transmission. Or they seem to be based on questionable assumptions about sewage and disease.

This was demonstrated in California when a university professor sought the basis for certain state water quality standards, and found more often than not that a standard had no firm foundation other than having been prescribed by "technical personnel" or "a com-

mittee of representatives of the interested departments." In another instance, the California Bureau of
Sanitary Engineering couldn't find out why the state
had settled on 500 milligrams per liter as the recommended top limit for the total dissolved solid materials
in drinking water. Why wasn't it 400, or 572, or some
other number? No one could say. A well-known sanitary engineer, P. H. McGauhey, commented, "While
rational men might agree that it is a reasonable standard, the search revealed that the mind of man does not
recall, nor do his records reveal, its origin. Most likely
it represents a value widely attainable in an America
much less thickly populated than today."

As Santee, California, officials learned when they
decided to reclaim sewage effluent for swimming,
tests for determining the health hazards of water
were far from refined. In fact the most widely used
test of all is a gross analysis which simply determines
whether or not water contains organisms found in
the intestines of warm-blooded animals.

It's a fast, inexpensive count of the coliform group
of bacteria, which are ordinarily harmless microbes
found in great numbers in the feces and urine of
warm-blooded animals. It's not a direct search for
pathogens (disease-causing microbes). The count only
indicates the degree to which the water has been
associated with excrement from humans, animals,
amphibians, and birds. If too great, the water is assumed potentially dangerous, because the probability
of its having acquired pathogens is higher.

Under United States standards water is considered
safe to drink if a certain average coliform count shows

that the density of such microbes is only one or less per 100 milliliters. This is not to say that water with a count of 200 or even 5,000 coliforms would necessarily harm the person drinking it. The higher counts would only indicate a greater probability of its being unsafe.

To discover exactly what pathogens a given body of water contains would entail a complex, expensive search of the water. An example of the problem was the expensive, difficult attempt to determine the presence or absence of viral organisms at Santee — although detection methods have improved since that time.

A more certain test of what harmful pathogens a specific body of water contains can be made — slowly and after the fact — by epidemiological studies. In short, people who have been in contact with the water are surveyed to see if it affected their health. Only a few such studies have been made, usually with people who have used certain bathing beaches, and they've often indicated that high coliform counts are not necessarily proof that water is loaded with harmful pathogens.

One of the most famous surveys was made in the late 1950's in England by the Committee on Bathing Beach Contamination. For five years the committee made bacteriological and epidemiological studies at forty bathing beaches, most of which were subject to sewage contamination. Coliform counts of the water ranged from 40 microbes per 100 milliliters to 25,000. While these counts indicated serious potential for harm to humans, the epidemiological studies showed that in truth the risk had been negligible.

Earlier, in America, the Joint Committee on Bathing Places of the American Public Health Association and the Conference of State Sanitary Engineers decided that they really couldn't establish standards that would distinguish between safe and unsafe water for swimming. Actually such standards are left up to state health departments. In the 1960's the state limits on coliform counts for safe swimming ranged from 50 to 3,000 coliforms per 100 milliliters.

But how can people swim with little health risk in sewage-contaminated water with coliform counts hundreds or thousands of times above drinking water standards? The answer doesn't lie in the counts, which are certain to be high if water is at all associated with excrement from warm-blooded animals; it's found in the community where the sewage originates. If the community is healthy, its excrement will be healthy — thus low on disease-causing microbes. If a community has a lot of disease, its sewage is more likely to contain a proportionate number of pathogens.

This point was stressed in a paper by Professor Melvin A. Bernarde of the Department of Community Medicine, Hahneman Medical College and Hospital of Philadelphia. In an appraisal of the health effects of applying sewage effluent to the land, Professor Bernarde discussed the decline of such diseases as brucellosis, typhoid fever, viral hepatitis, paralytic poliomyelitis, salmonellosis, shigellosis, and trichinosis, and then commented: "From these trend lines, it is evident that the potential for dissemination of those viral and bacterial diseases transmitted via fecal matter, is low indeed. What these lines tell us, is that although even raw fecal matter might be broadcast upon the land, relatively few infectious agents would be broadcast along with it. Furthermore, one could reasonably expect that given adequate treatment, the ... effluent could be expected to contain orders of magnitude fewer viable pathogenic agents."

Going back to the chapter's beginning and the first of our two questions, the answer seems to be that while the health hazards of sewage must always be approached with a cautious, conservative turn of mind, they are not as serious in today's healthy communities as when infectious diseases were more prevalent. Furthermore, sewage treatment, because of its unfavorable environment and the time it takes, leaves the effluent with far fewer microbes than when it was raw waste. And, finally, disinfecting effluent with chlorine reduces the organisms even more. So the health hazards of chlorinated sewage effluent applied to land can usually be considered negligible.

But now if we answer the second of our two questions we find that effluent applied to the land poses even less of a threat to health—indeed far less than does the common practice of dumping sewage, raw or at various levels of purification, into rivers, streams, and lakes used by millions and millions of people.

Nature's purification processes using soil, air, and sunshine are remarkably efficient at rendering contaminated water harmless. People in ages past in search of clean water learned this truth from experience, and today it's an article of faith with nearly everyone. It's also backed up by numerous scientific investigations that have tested the survival of disease-causing organisms in soil and on plants under conditions far more severe than would occur in the conventional land treatment of secondary effluent.

First of all, there's plenty of evidence that soil is a generally powerful agent for removing hazardous microbes from waste water. We've already learned of an outstanding example of the reduction of bacteria and viruses in the soil percolation beds at the Santee and Whittier Narrows sewage reclamation projects. The same capability has been demonstrated in other cases. For instance, studies of army latrines dug from the earth have demonstrated that bacterial contaminants are removed from water as it travels laterally through the soil. The efficiency depended on the nature of the soil and the speed with which the water moved. One such study concluded that water contaminated by a latrine would be safe to drink after eight days' travel through the earth. Depending on the soil's porosity, this travel time could represent anywhere from a few feet to a couple of hundred feet.

One of the most quoted studies of how soil can remove bacteria from waste water was conducted at Lodi, California, in the early 1950's by three University of California scientists, Arnold E. Greenberg, Harold B. Gotaas, and Jerome F. Thomas. They tested both primary and secondary effluent from the Lodi municipal sewage treatment system. It was applied to the ground surface in several circular basins enclosed by metal dikes, each nineteen feet in diameter. In four of the basins dry wells were sunk to a depth of thirteen feet. At several intervals along the vertical sides, collection pans gathered samples of the effluent as it infiltrated down into the earth, a "Hanford fine, sandy loam." After analysis of hundreds of samples over twenty-eight months, the California scientists concluded: "A bacteriologically safe water can be produced from settled sewage [primary effluent] or from final effluent [secondary] if the liquid passes through at least four feet of soil."

Of course all the microbes that might be brought to a land treatment system would not suffer soil filtration. Some would remain on the surface, especially on plants which might be harvested and removed from the site. The question, therefore, arises as to whether these microorganisms might somehow endanger the health of animals or humans — even though, in the case of the latter, no one proposes (nor do most state laws allow) that such crops be used for direct human consumption. The answer is that the threat is negligible, if it exists at all. This is revealed by practice on the Paris, France, sewage farms, and it is backed up by some long-standing studies conducted by three investigators of the New Jersey Agricultural Experiment Station at Rutgers University: Willem Rudolfs, Lloyd L. Falk, and Robert A. Ragotzkie.

Their investigations, reported in 1950 and '51, were supported by the U.S. Army Quartermaster Corps for some very practical reasons. With troops spread all over the world, the corps had to obtain a great deal of fresh, raw food grown in areas where human excreta was used for fertilizer, directly as "night soil" or less directly as raw sewage or highly polluted river water. Army officials, who had always warned against eating raw foods in such areas, still had a number of questions as to the real nature of the hazards. In a long and thorough study, the Rutgers scientists tried to find out.

They grew tomatoes, lettuce, spinach, and carrots irrigated with sewage. To make sure that the plants were truly exposed to pathogenic organisms, they artificially contaminated them with liquids carrying raw feces and quantities of microbes capable of producing a low grade of typhoid fever and forms of dysentery. They also included helminth eggs associated with intestinal worms. In other words, if any plants were ever to qualify as dangerous to human health because of sewage contamination, the Rutgers vegetables would lead the list. Then the scientists studied the plants to see what nature would do to the contaminants. Several of their conclusions were forceful testimony about the power of air, sunshine, and soil for rapidly reducing serious health hazards. Here are the first four:

"1. No evidence has been found that pollutional bacteria, amoeba, or helminth eggs penetrate healthy surfaces of vegetables or cause internal contamination.

"2. Vegetables to be eaten raw can be grown without health hazard in soils that have been subject to sewage irrigation, night soil application, or polluted stream water irrigation in years prior to the season in which the vegetables are grown.

"3. Vegetables grown under conditions of surface irrigation show no higher coliform concentrations than those grown on normally farmed soil, whether sewage was applied before the plants were set or while the plants were growing.

"4. If sewage sludge or night soil [raw feces] are applied on the soil surface, or sewage effluents are applied by overhead irrigation during growth of vegetables, such an application should be stopped at least one month before harvest. If this precaution is taken, the crop will show no higher bacterial contamination than when farmyard manure or artificial fertilizers are applied."

The remaining nine conclusions further supported the contention that nature will eliminate the health hazards of contaminated plants in a reasonably short time. For example, the microbes responsible for the low-grade typhoid and one form of dysentery failed to survive on vegetable surfaces for more than a week. Another microbe causing dysentery lasted less than three days in dry periods. The helminth eggs degenerated within thirty days, and no completely developed eggs were found on the contaminated plants. On the other hand, the harmless coliform bacteria always seemed present on the plants, sewage contamination or not. Thus any conclusions based on the common coliform count would be meaningless.

These classic Rutgers studies and others since have established that dangerous organisms applied to soil and plants face a losing battle with the forces of nature. But one more question always arises when spray irrigation methods are being considered. Is it possible that hazardous pathogens in aerosols (extremely fine droplets) may be carried great distances from the sprays by the wind?

The question has led to numerous studies, mostly concerned with aerosols created by the vigorously churning and splashing tanks of conventional sewage treatment plants. First of all, the research has confirmed that the tiny droplets can hold microbes. But how far they can travel on the wind is pretty much decided by how long it takes before the aerosols dry up, thus leaving the living organisms to the same fate, a death technically called "desiccation." The studies have indicated that they ordinarily desiccate in a few seconds, although some resistant strains of bacteria may last somewhat longer. The time is naturally related to how long an aerosol itself survives in the moving air, and this is a function of droplet size, wind velocity, relative humidity, temperature, and sunlight.

While the danger of aerosols is apparently not great, it is one that can't be ignored, and precautions are usually advised. Chlorination to disinfect the sprayed effluent is a first step. Spray equipment designed to reduce the aerosol effect—using low pressure with sprays directed downward—is another remedy. Storage in the system, which eliminates the need to irrigate during periods of high wind velocity, high dew point, and low temperatures, provides further safeguards.

In discussing the two questions posed at the chapter's beginning, we've confined the health topic to infectious diseases. We should also consider the chemical hazards in sewage where there is possible contamination of groundwater supplies. Scientists are not completely sure, for example, about the absorption of the various heavy metals by soil and plants. As we learned earlier, some of these ingredients in trace quantities are essential to plant growth, but in larger, toxic amounts they can be serious threats to the natural biological processes so crucial to water purification. Until we know more, the safe solution is to prevent toxic amounts of such substances from getting into sewage at their sources, often as industrial waste. Meanwhile scientists who are studying the living filter hope to determine more precisely the amounts of heavy metals that various soils and plants can be expected to take up.

One sewage ingredient widely discussed in terms of health is nitrogen. As we've seen, it may slip through soil (where it becomes nitrate) under certain circumstances. Although sewage is not necessarily the main source, the danger of undue nitrate accumulations in the nation's groundwater greatly concerns health authorities, environmentalists, and others. It is of such importance that the National Academy of Sciences issued a special 106-page report in 1972, entitled *Accumulation of Nitrate*.

The health problem centers on Methemoglobinemia, a long name for the affliction commonly called "blue baby." It was determined in 1944 that the problem arises from nitrite, which is derived from nitrate, and in turn from nitrogen. The damage occurs when oxygen transport is impaired in a baby's blood. As a preventive measure, the U.S. Public Health Service has set a standard that drinking water should contain no more than 10 parts per million of nitrogen as nitrates. The limit recommended by the World Health Organization for Europe is 22.6 parts per million. As this disparity indicates, there are still unresolved questions in official minds. Despite all the talk about the difficulty, in the dozen years prior to 1972 there were only two blue baby cases in the United States — one in Illinois, one in Colorado, both the result of the mothers' boiling water too long and concentrating the nitrite content.

Nevertheless, there is an intense search for the cause of undue nitrate accumulations in this country's groundwaters. For example, in 1972 California scientists began drilling four hundred test wells in the state to study the problem. The main sources under surveillance in the United States include:

— municipal and industrial waste water dumped into the nation's waterways;

— animal feed lots and poultry farms where huge quantities of nitrogen-rich manure washed by rain may pollute groundwater;

— internal combustion engines, especially automobiles, and coal-burning furnaces emitting nitrogen oxides that eventually get into the earth and water;

— agricultural runoff to surface water and leaching to groundwater, especially from farms using nitrogen-rich artificial fertilizers;

— and septic tanks and cesspools which take care of 25 percent of this country's domestic wastes.

Despite all these potential sources, the National Academy of Sciences had to report that its study committee "found an appalling lack of information about the significance of the various sources" and how the problem — which isn't easily defined — can be controlled.

Nevertheless, sewage is a source, and its treatment must be concerned with nitrogen removal in consideration of the possible health hazard. Conventional sewage treatment systems discharging nitrogen-laden effluent into our waterways can hardly help the problem. Nitrogen removal in the tertiary treatment systems now being proposed is one of the most difficult, undependable parts of the process. But the well-designed living filter, using nature's way of removing nitrogen by plants and microorganisms, already has proven that it can provide the answer.

Indeed, with any of the health problems that we have discussed, one can safely say that our conventional sewage treatment and disposal practices are pretty certain to create more, not less, of a threat to the public health than will the intelligent use of nature's processes, which have proven over the eons that they can truly reduce the hazards.

18

Two Renowned Research Projects

Anyone seriously interested in land treatment, especially a scientist or engineer, is certain to learn about, and perhaps visit, one of at least two meccas in the United States.

The most renowned is at Pennsylvania State University in the Nittany Valley along the western slopes of the Appalachians. Here several scientists and engineers with different specialties have established a land treatment system using secondary-treated sewage from their community, State College, Pennsylvania. It operates year round in a relatively wet, varied climate, including a cold, snowy winter season when fields can freeze concrete-hard. The project demonstrates how land treatment can serve as a tertiary treatment system that at once fully completes the job of sewage treatment, controls water pollution caused by secondary sewage effluent, maintains precious open space, utilizes otherwise wasted nutrients that could help pay for the project, and returns clean, fresh water to the underground storage areas from which it came, as opposed to losing it by discharge in a local stream.

The second project, which primarily attracts scientists and engineers, is on the desert immediately outside Phoenix, Arizona. Using secondary sewage effluent from the city, it is a pilot land treatment project with the provocative name, "Flushing Meadows." The project, which is operated under the U.S. Department of Agriculture Soil and Water Conservation Laboratory, is significant because it may open the way to purifying extra large amounts of waste water on relatively little land. Flushing Meadows has special interest to the many experts concerned over nitrogen removal where sufficient land with crops might not be available.

Let's consider each project more closely.

Back in the 1950's State College, Pennsylvania, dominated by the large and expanding Penn State University, began seriously polluting nearby Spring Creek with secondary sewage effluent from a combined university and town treatment plant. Below the area where the effluent entered the creek the once clear water was choking up with algae, and the perennial trout died off to be replaced by unsavory suckers. Upstream Spring Creek remained, as always, a beautiful, clear trout stream. At first, the town considered controlling the pollution by piping the effluent considerably downstream to where there was enough water to dilute the effluent—at least for a while. But this idea was only temporary buck passing, not the permanent solution that people wanted.

After due consideration, a group of Penn State scientists and engineers decided the answer might be land treatment, but at the time they could find no good precedent for a municipal system like the Nittany Valley in terrain and climate. Still there was a lot of enthusiasm to set a precedent, and around 1961 they began, slowly and meticulously, to design what they eventually called a "living filter" for State College.

The group, including agricultural and civil engineers, agronomists, biochemists, foresters, geologists, micro-biologists, and zoologists, started extensive research on some university land four or five miles from the campus. First they studied the earth from the topsoil to the underlying geology and hydrologic conditions to determine exactly where land treatment might work best. The group then began a five-year experiment to test the practicality of using the land for tertiary treat-ment of the community's sewage. From its inception numerous monitoring programs were established to record the changes the effluent irrigation would cause in the land, the surface water, and the groundwater. For example, the investigators selected a series of private wells for homes in the immediate vicinity, and with the owners' permissions instituted a long-term monitoring program of the water supplies.

In May 1963 chlorinated secondary effluent was pumped through a pipeline to the experiment site for spray irrigation at varying rates on different test plots of both agricultural and forest land. Other comparable plots, "control plots," were not sprayed, so as to compare the irrigated with nonirrigated land. In the next five years the research developed an unprecedented amount of scientific and technical data on all phases of land treatment.

The scientists were most astonished by the remarkable crop yields produced by the sewage nutrients. For example, their control plots fertilized with normally recommended commercial fertilizers were yielding 2.27 tons of alfalfa hay per acre, but a comparable acre receiving the equivalent of one inch of sewage effluent per week produced 4.67 tons, and an acre receiving two inches yielded 5.42 tons. Increases of the same magnitudes were evident in other crops. The unsprayed acre with commercial fertilizer yielded only 63.30 bushels of corn, compared to 114.40 from the acre receiving a weekly, 1-inch dousing of effluent— which, the scientists figured out, was comparable to applying 1,000 pounds of fertilizer per acre. They also discovered that with some crops too much effluent could retard the yield.

A series of test and control plots was also established in wooded areas to study the growth rates of sprayed and unsprayed trees. The most dramatic results occurred with white spruce saplings. After receiving effluent for five years they were almost twice as high as those in the control plots. Also, the ground vegetation in the sprayed areas was double that of the control land. In another sprayed area the trunk diameters of young, maturing oaks increased from 27 to 83 percent over unsprayed oaks. Some trees didn't respond to the effluent, and in some, growth was re-tarded, as in the case of red pines which thrive on dry soil. But enough varieties—like white pine, Norway spruce, European larch, and Japanese larch—did respond, revealing that they were using up sewage nutrients.

All during the spraying the living filter's capability for renovating waste water was analyzed. In the test plots water samples were drawn from a few inches be-low the soil surface to many feet, down to the ground-water. They were tested to determine what the soil and plants were removing. Also, groundwater samples from many domestic wells, natural springs, and test wells around the experiment site were analyzed. And nutrient removal in the agricultural plots was ascer-tained by checking the plants for minerals absorbed from the effluent.

The living filter was best at removing phosphorus, both in the agricultural and forested plots. By the time the sewage water had descended to the soil's four-foot level, the plants and earth had removed nearly 100 percent of the phosphorus from the effluent. Ground-water tests from both deep and shallow wells on the project site showed only .04 milligrams of phos-phorus per liter of water—which was about the same as in the unpolluted sections of Spring Creek. The creek's polluted water had from 2.3 to 3.5 milli-grams of phosphorus per liter.

The living filter also scored high in removing other nutrients and micronutrients of sewage, such as po-tassium, calcium, magnesium, and sodium, but the opposite was true of nitrogen. To prevent it (in the form of nitrate) from reaching the groundwater, the scientists needed a sophisticated understanding of how the soil and plants could help. The scientists had to find what crops used the most nitrogen and at what time during growth. Timetables based on this knowledge were worked out for applying effluent at controlled rates that would use the area's croplands, grasslands, and forest lands to the greatest advantage. However, the project leaders concluded that the living filter met its most difficult test in nitrogen removal. It required close, intelligent control all the time.

They were especially surprised and pleased by one result. When the experiment began, people were still using the so-called hard detergents with alkyl benzene sulfonate (ABS), a substance that was accused of seriously polluting water. The Penn State scientists found their soil removing up to 90, even 99, percent of the ABS in the effluent. (ABS is greatly reduced in today's "soft detergents.")

Samples of groundwater from under and around a wide area of the experiment were constantly tested with the common coliform count to determine how bacteria fared on a descent through the soil. In all the tests the microorganisms were nearly missing after only a few feet of infiltration. Frequent samples would show no coliforms at all. Some of the project directors demonstrated supreme confidence in soil filtration when they drank water drawn from just under the earth immediately after the ground had been sprayed with sewage effluent.

As the experiments succeeded, the university scientists and engineers expanded the project to make it a practical system for the town. One of the problems was how to proceed in the winter. They considered using storage lagoons to collect effluent during cold months and spraying only in the warmer seasons, but the group's geologists questioned the advisability of constructing lagoons on the Pennsylvania land. Instead, the project leaders decided to spray during the winter in their forest tracts, taking advantage of the porous forest floor. University engineers developed spray equipment that would not clog by freezing, and the forest irrigation proved it could cope with whatever the Pennsylvania winter offered.

By the 1970's the Penn State project was treating an increasing share of the community's sewage effluent, and there were plans to provide land treatment for the entire discharge of some four million gallons a day.

In the meantime thousands of people from all over the world were visiting this mecca which was publicized in many ways, from university films to technical journals to newspaper stories. The visitors included sanitation officials, of course, but also lay citizens, from fishermen to conservationists, who doubted that conventional sewage treatment would ever control water pollution.

Many of the visitors asked how their communities could take advantage of the living filter. Project leaders, like Louis T. Kardos, William Sopper, and Earl Meyer, were always hesitant in answering. Their research and experience had taught them that the Penn State results could not be generalized. Another geographic area with different climate, soil, and geological conditions

might require a different program from what had succeeded in the Nittany Valley. With that understood,
these experts on the living filter were glad to relate
their work to other locales.

In the Penn State system the croplands and woodlands serve best with an average 2-inch application
of effluent per week, which means that the waste
water from about 10,000 people, or 1 million gallons
per day, can be treated with about 129 acres of land.
Thus a small city of 100,000 citizens would require
about 1,300 acres.

The cost in other communities would be something
the Penn Staters couldn't predict. It depends on too
many variables, from land costs to possible returns on
crop yields. But they have been glad to point out the
potential benefits, from preserving open space to
dealing with water pollution fully and finally.

One of their biggest benefits was the return of substantial volumes of good water to the natural supply,
which had suffered ominous drops in the water table.
Instead of abandoning State College's used water
down Spring Creek and eventually to the ocean via
Chesapeake Bay, it now went right back to where it
originated. Analyses of the wells supplying the university community clearly showed that they now
contained recycled sewage water purified to nature's
demanding standards.

But with all its plus marks the Penn State system is
limited to applying a relatively small amount of effluent per acre per week. Increasing the weekly application rate could, of course, reduce land needs, but it
would also raise the potential for too much nitrogen-

nitrate to slip through the living filter. How to beat the nitrogen problem is thus a key question in land treatment. For this reason people who are sold on land treatment frequently visit the Phoenix, Arizona, research project.

Instead of meeting a group of scientists in a busy university setting, visitors to the Flushing Meadows project are usually driven by a soft-spoken Dutch scientist to a fenced-off area of the desert with only a trailer as a field laboratory. The scientist is Herman Bouwer who comes from Holland, where he specialized in reclaiming land from the sea and thereby learned a great deal about the movement of water under ground. Under the U.S. Department of Agriculture he is now responsible for the Flushing Meadows project, which started in 1967.

The project was initiated as a pilot study seeking a solution to a prime problem of the Salt River Valley, where Phoenix is located. About a third of the municipal and agricultural water supply is obtained by wells from a dwindling groundwater supply. For over a decade the water table dropped around ten feet per year because of the valley's increasing population. In the 1960's when the successful Santee, California, project received a lot of publicity, Phoenix officials were prompted to consider renovating and reusing the millions of gallons of waste water escaping daily from their community. This led to the desert experiment conducted by Bouwer.

The Flushing Meadows project goal was more like Santee's than Penn State's. It aimed to reclaim waste water for recreation, agriculture, and possibly certain

industrial applications, rather than allowing it to
return to the groundwater for reuse in the local domestic water supply. As was true at Santee, the earth
would serve as the renovating agent, with the purified
water being captured for reuse before it mixed with
the natural underground supply. But at Phoenix the
volumes of waste water far exceeded Santee's, and
this led to an effort to renovate as much effluent as
possible on the least amount of land. The problem
came back to the old challenge of nitrogen removal.

The pilot project was set up just west of the city
outskirts on one side of the dry, half-mile-wide bed of
the Salt River. Six parallel infiltration basins were cut
in the earth's surface, shallow pans 700 feet long and
20 feet wide. Their natural bottoms consisted of about
3 feet of fine, loamy sand on some 240 feet of coarse
sand and gravel, leading down to a solid layer of clay.
Eight test wells were drilled to varying depths around
the basins' immediate area to allow for sampling water
as it percolated down through the soil.

When Bouwer was ready to use the basins, secondary sewage effluent from a city-activated sludge plant
was pumped to the site, and the Dutch scientist began
a long series of experiments which he hoped would
reveal how the soil might be used to purify large
volumes of waste water, at least to recreational standards.

Where Penn State was applying only two or three
inches of effluent a week, Bouwer began inundating
his basins with two or three feet per day. He soon
confirmed what everyone quickly learns, that soil only
a few feet deep is fully capable of removing bacteria,

phosphorus, and numerous other ingredients already discussed, but not the troublesome nitrogen-nitrate. For the next several years Bouwer devoted most of his research to this problem.

While Penn State solved the problem with crops, Bouwer realized that this was not his solution. The sprays in Pennsylvania were adding relatively few pounds of nitrogen to the soil, quantities that crops could absorb. But Bouwer's voluminous applications meant he was literally putting a dozen or more tons of nitrate a year onto his basins, quantities that crops couldn't even begin to absorb.

The Phoenix scientist tried an entirely different approach to the problem. He would have the nitrogen-nitrate transformed by soil bacteria to a gaseous form, and released to the atmosphere. The process, "denitrification," works like this:

In soil under aerobic conditions bacteria depend upon atmospheric oxygen to digest the earth's organic material. If that oxygen is cut off, creating an anaerobic condition, the bacteria may obtain oxygen from nitrate, if it's present in the soil. Removal of the oxygen leaves free nitrogen gas which escapes to the atmosphere.

To activate this process Bouwer found he could turn the soil in his infiltration basins anaerobic by completely inundating them with effluent, thus shutting off oxygen from the air. Soil bacteria then took oxygen from the nitrate which had already been converted from the nitrogen of the waste water. After this denitrification had worked for a while Bouwer stopped applying effluent and allowed the beds to dry out. Meanwhile the nitrogen, now in gaseous form, escaped into the air.

The efficiency pretty well depended on how much effluent the scientist applied for how long, and the length of the drying period between inundations. Then he discovered he could enhance denitrification by planting Bermuda grass on the surface of his beds. The grass roots helped deplete the soil of oxygen, thus fostering the anaerobic condition, and they added organic material to the earth, which stimulated the key bacterial action for denitrification.

How to use the process for maximum nitrogen removal eventually became the focus of Bouwer's research. He could only proceed by trial and error, inundating and drying out the basins hundreds of times with various quantities of effluent over different periods of time, all in search of the optimum conditions. The scientist once compared the pursuit to "being blindfolded in a dark cellar trying to catch a black cat."

In five years of such tests Bouwer found that over a long period he could remove an average 30 percent of the effluent's nitrogen, but often the removal was as high as 90 percent. The average was dragged down by brief peak periods when the removal process worked at its lowest level. With further improvement of his techniques, the scientist was confident he could reduce the peaks and raise the average nitrogen removal much closer to the 90 percent level.

By the early 1970's the Flushing Meadows pilot project had confirmed that a large-scale sewage reclamation system could be designed for the 80 million gallons of effluent per day discharged from Phoenix. Using Bouwer's technique it would require only three hundred acres of infiltration basins. They would be situated at one side of the broad, dry bed of the Salt River, and would be designed to receive some three hundred feet of effluent per year. The renovated water would be pumped from strategically located wells at the center of the river bed. Bouwer assured community officials that the billions of gallons of renovated water annually would be pure enough for recreation that included swimming, all kinds of irrigation, and most industrial uses. The cost estimates for infiltrating the water and pumping it back to the surface was set at about fifteen dollars a million gallons.

The promise of such a water bonus in the dangerously dry Phoenix area was delightful to hear, and by 1973 community leaders were planning to use the welcomed deluge for numerous purposes. For example, they were considering a greenbelt recreational area comparable to the one at Santee, but far more extensive. With the promise of sewage reclamation, they could now think of using water without the continual, nagging fear of overtaxing the precious groundwater supply.

The two land treatment research projects, so different in climate, soil, and community demands, illustrate how nature is ready to cooperate with our onerous sewage problems when we make the effort to understand and use her astonishing purification systems intelligently. To scientists the Penn State and Phoenix projects make a point that they feel has to be stressed over and over. There will be no cut-and-dried land treatment systems. Each one has to be customized to the particular local demands of nature and the people to be served.

19

Down to the Farm
With Big City Sludge

Most of the curses from city sewage men have been reserved not for the millions of gallons of effluent which disappear down the river but for the tons of thick, gooey sludge which stay and literally threaten their burial. Indeed, some authorities have surmised that many sewage plant operators have been seriously distracted from other problems, like water pollution, by their sludge difficulties.

The decade of the 1960's with its culmination of urban growth, brought the problem of big city sludge disposal to a crisis that forced new approaches. The leader in new thinking was the Metropolitan Sanitary District of Greater Chicago, where sludge disposal took 46 percent of an annual $15 million operations and maintenance budget. Getting rid of some nine hundred tons of sludge per day was the district's largest single sewage treatment problem. How it was solved has significance for everyone interested in the environmental impact of sewage. The leaders behind the effort were the district's president John Egan, General Superintendent Bart Lynam, and Chief Engineer Forrest Neil.

In the 1960's Chicago was using several methods for "digesting" the daily sea of sludge produced by numerous sewage plants, including the world's largest, the West-Southwest plant at Stickney, capable of secondary treatment for over a billion gallons of sewage a day. Essentially the sludge was heated and "stabilized," so as to reduce the biological activity to eliminate odors and other problems. It still remained wet and thick, the consistency of pancake batter (only black), and the processing operation had a big drawback.

As the quantities of sludge increased, digesting operations became serious air polluters. On still days smoke from two giant stacks at Stickney sometimes settled and stopped traffic on nearby highways. At times racetrack patrons two miles away claimed the smoke prevented their seeing the finish line. Scrubbers were installed in the smokestacks to reduce air pollution. They demanded three thousand gallons of water for every thousand gallons of processed sludge — and they didn't work. This and other efforts to deal with the sludge were not only expensive, they were dangerous to personnel. In one operation there were four deaths.

And then, no matter what the Chicago sanitarians

did, they still faced the problem of getting rid of the digested sludge. It was generally pumped into large, unattractive storage lagoons near the plants. There in time the solid materials settled and the remaining liquid was drawn off and treated again. Eventually each lagoon filled up with solids, and Chicago used up space at the rate of about 350,000 cubic yards a year. By the 1960's land was simply running out, and the district's expensive disposal efforts had no future. At one point a quarter million dollars were spent to clean out the overflowing lagoons of one treatment plant. When the project was finished, the moving contractor had simply transferred 200,000 cubic yards of waste material to another hole in the ground a mile and a half away.

The same story, with minor variations, is being played out in many big cities. Even coastal cities, where the ocean seems like a bottomless sink for sludge disposal, face trouble. They are finding that the sea is pollutable and sludge dumping has to cease.

Thinking as hard as possible, Chicago's sanitation officials kept coming back to an old alternative. Why couldn't the sludge go downstate to rural Illinois? But downstate citizens traditionally opposed the idea. Their objections even had political overtones. The heavily Republican rural counties said they'd never take Cook County's Democratic excrement.

Nevertheless, Chicagoans had to overcome the resistance. And the only way was to prove that the waste was literally black gold, fertilizer which downstate farmers couldn't refuse, especially when the offering was free.

For decades, ever since conventional treatment plants were first introduced, operators had recognized that sludge, like sewage effluent, had great potential for stimulating green growth. They couldn't help but see how healthy, handsome tomatoes sprang up out of dried sludge piles from seeds that had found their way through sewage systems from the users' homes. The black material was also fabulous for improving grass around the treatment plant or on the operators' lawns at home.

Over the years technical journals frequently told of how some enterprising plant operator was putting his sludge to work in the community or on nearby farms. In the 1940's, for instance, an operator named Damoose in Battle Creek, Michigan, used a six-hundred-gallon tank truck with a hose to spray local lawns with sludge as a soil conditioner. Assuring customers that he brought no pathogens, he charged ten dollars for the service, which included watering the sludge into the soil to prevent odors. Damoose's lawn service became so popular that he had to increase the size of his tank truck. He also claimed that it was all good public relations for the sanitation department.

Most such ventures were based only on practical experience, but at a few places research defined some of the good and bad growth features of sludge. Work at Milwaukee in the 1920's led to Milorganite, the dried sludge product discussed in Chapter 14. Around then wet and dried sludge was compared with stable manure and commercial 5-8-5 fertilizer in an experiment near Baltimore. The sewage materials proved best for potatoes, cabbage, and sugar corn, but not so good for spinach and tomatoes. Henceforth a number of other experiments pointed up the value of sludge as both a soil conditioner and fertilizer that had roughly the same nutrients as sewage effluent—the main exception being a lack of potassium. But over the years, despite the evidence of its beneficial qualities, the application of sludge to land was pretty much confined to rural towns and small cities. Big city plants, often engulfed in urban spread, followed other avenues of sludge processing and disposal comparable to those that were leading Chicago down a dead-end street in the 1960's. Many tried incineration, which was high in cost and increased air pollution.

Before the Chicagoans could go to the Illinois farm country with their sludge, they recognized a need for much more research to prove definitively that their black batter was safe and truly worth the name of gold. Around 1967 the Metropolitan Sanitary District began a series of scientifically based programs to test sludge and demonstrate that it could be used as a good, safe fertilizer to the benefit of farmer, city dweller, and all ecologically concerned citizens.

One of the most important programs was conducted with the Department of Agronomy of the University of Illinois. It was a comprehensive experimental program directed by agronomist Thomas Hinesly who worked in the laboratory, in greenhouses, and in the field to ascertain a number of facts about the use of sludge on soil. The Hinesly study was, for example, designed to learn about the best time and frequency for applying digested sludge to land, how crops would respond, and the possible hazards of contaminating soil, water, and crops with chemicals or bacteria.

The researchers worked at a farm site near Elwood, Illinois, where they established a series of plots, ten-by-fifty feet each, separated from one another by plastic moisture barriers. Here the investigators meticulously studied the effects of applying Chicago sludge to native soil with a wide variety of crops, like alfalfa, corn, kenaf, Reed canary grass, and soybeans. At the same time a greenhouse study looked into soybeans grown with exceptionally large applications of sludge.

Soon the research confirmed that the digested sludge was a rich source of nitrogen, phosphorus, and micronutrients, which, in its partially dried state, still contained a valuable quantity of water. The study developed considerable data on how to use the material without producing undue nitrogen-nitrate in the drainage water. For instance, the scientists learned that simply storing sludge for a certain time in a lagoon reduced the nitrogen content (by discharge to the atmosphere) so that more of the material could be safely applied to the soil. And the study showed that coliform bacteria decreased as soon as digested sludge was applied to the earth. Again, lagoon storage could reduce coliform organisms even more—for that matter, to a safe level before application of the sludge to the soil. Further study provided good evidence that viral organisms could not survive the heated digestion process at the sewage plant.

Above all, crop yield data was exciting, especially for Illinois corn farmers. Comparing corn crops showed that an acre of land receiving a half-inch of sludge a week produced 114 bushels, while an un-sludged acre offered only 66 bushels.

In 1971 Hinesly stated: "We can utilize the sludge anywhere crops can be grown and without regard to soil types. It can be used on sandy or clay soils to improve their capabilities to hold water. The sludge furnishes a very good source of slowly released nitrogen and phosphorus for plant growth, and the organic matter improves tilth [good, cultivated land] and furnishes a root bed for crops."

He also pointed out that the batterlike sludge offered farmers another great advantage — it was easy to apply compared to dry, bulk fertilizers. Farmers could spray the material on the land with conventional irrigation rigs.

Another sludge experiment program initiated by the sanitation district provided a dramatic demonstration that sludge-stimulated crops could make good neighbors. Next to the district's Hanover plant, a few yards from a large concentration of suburban homes, fields were divided into six plots, a little more than an acre each, and they were fertilized with digested sludge from the sewage plant. Experimental corn crops were raised with outstanding yields without detrimental effect to the soil, crops, or neighborhood. The program was so successful that the farm site was extended into a few hundred nearby acres where houses had not been built.

One of the most astonishing uses of digested sludge revealed it could support crops on a desolate waste-land of fine sand and ground glass. The Sanitary District engineers worked with the Libby-Owen-Ford Company at Ottawa, Illinois, to reclaim some twenty-two acres of dried-out lagoons where the company had dumped waste materials from the grinding and polishing of automobile windows and windshields. Primarily the waste material consisted of 76 percent grinding sand and 17.5 percent finely ground glass. In the fall and winter, prairie winds swept the fine materials around Ottawa, to the citizens' great discomfort. In particular, the spray created a visibility hazard on a nearby state highway.

Barges, each carrying 1,200 tons, transported digested sludge from Chicago to Ottawa, where it was applied to the desolate waste beds at the rate of 170 tons per acre. Three kinds of grass (rye, orchard, and brome) were planted, resulting in a dense cover of vegetation in three months. The disturbing dust problem was solved.

A report on the project concluded: "The sludge supplied an abundance of all the essential plant nutrients and copious amounts of organic matter. Applying sludge dramatically reduced the sodium concentration of the soil [grinding wastes], the main deterrent to plant growth. The project showed that sludge application can reclaim sterile land in less than one year."

While these and other projects were proving the value and safety of applying sludge to the soil, the Metropolitan Sanitation District contracted with a newly formed company, Soil Enrichment Materials Corporation (SEMCO), established to haul sludge by railroad tank cars to Illinois farm communities. The first recipient was Arcola, Illinois, where the new company had bought a 155-acre demonstration farm and had made arrangements to apply sludge without charge on the land of local farmers willing to co-operate.

But before the first waste material was hauled into the area, SEMCO and the Chicago Sanitation District conducted an intensive public education program to allay the old fears about city wastes being dumped on the country folk. For example, a group of Arcolans, civic leaders, farmers, realtors, newspaper publishers, bankers, and concerned citizens—even the executive of a commercial fertilizer company—were taken to Chicago for a grand tour of the mammoth Stickney plant and other points of interest related to the city's sludge production. The visitors heard speakers and saw demonstrations offering evidence that the sludge coming their way would enhance their land but do them no harm nor create any nuisance.

Soon after they returned home, trainloads of sludge, thirty-six tankers at a time, arrived continually with some three thousand tons of waste material a day. It was pumped into a thirty-five-acre storage lagoon, and in due time land application began with irrigation equipment. The bad smells that some citizens feared didn't arise. In fact, earlier one person had complained about the smell to Richard Williams, the Arcola newspaper publisher, who informed the complainant that the first trainload of sludge had yet to arrive in Arcola.

SEMCO continues to haul sludge into the Illinois farm country, although acceptance hasn't always been as smooth as in Arcola. In Grundy County the public education job was either insufficient or ineffective, and the sludge sprayers were temporarily stopped by a court injunction which was lifted upon appeal. Meanwhile, SEMCO, having learned that public knowledge is a key to such projects, firmed up its public communications by hiring John Lear who had long served as the highly regarded science editor of the *Saturday Review* magazine.

Perhaps the most rewarding aspect of the metropolitan district's sludge disposal program can be found some two hundred miles southwest of Chicago, where strip mining has ravaged Fulton County leaving the land virtually nonproductive. The program is called the "Prairie Plan." In 1970 the Chicago District began buying the scarred land (at some $350 per acre), and by 1973 it owned 13,000 acres. The plan called for restoring the bleak territory to productive land for farming, recreation, and other community uses.

Immediately the district constructed large holding lagoons, which would serve as part of the sludge digestion process, and the job of transporting the material from the windy city commenced with barges coming down the Illinois River. Meanwhile construction equipment prepared carefully designed application sites on the strip-mined land. They were planned so that runoff would not pollute the area's water. Essentially the design called for the development of a series of grassy fields surrounded with hedgerows. Stream banks were to be reforested. The roots of trees and hedgerows would serve to catch up nutrients that might move laterally through the soil with water. Also, the designers provided for water monitoring systems to give early warning of possible water pollution throughout the reclamation sites. The areas included reservoirs to which polluted water could be diverted, stored, and then recycled to the land for natural purification.

In 1971 sludge application began by using powerful mobile sprays to cover an initial eight hundred acres. In 1972 some of the newly covered land yielded sixty bushels of corn per acre on stalks eight feet high. The best that nearby, unsludged land could produce was thirteen bushels on sickly, three-foot stalks.

Meanwhile the Metropolitan Sanitary District was devoting about 60 percent of its sludge production to the Prairie Plan, and a lot of space remained in Fulton County where the land-enriching material would be welcomed far into the future. Some forty thousand strip-mined acres qualified for reclamation. By 1973 the district had spent $35 million establishing the Prairie Plan, and expected to use another $100 million for sludge transport and the purchase of more land. It was a lot of money simply to be rid of something, but compared to the past it was far less expensive in dollars, and it was ecologically sound. At the same time, the once negative downstaters were finally glad to accept Cook County waste regardless of its political affiliation. Fulton County land was on the way back up from damage that most people believed could never be repaired.

Chicago's success prompted other cities to think about the combined possibilities of sludge disposal and the reclamation of strip-mined land. According to Hinesly, existing unreclaimed land strip-mined for coal in the United States, plus that which will be ravaged by 1984, adds up to over a million acres. This could take a lot of sludge for productive use for many years.

A quarter million acres of unreclaimed mining lands exist in Pennsylvania alone. Around 1970 the Penn State scientists who were so interested in sewage began studying the effects of sludge on strip-mine spoil material trucked from stripped areas to the university farm. The material, which had remained barren for nearly a quarter of a century, was put in large boxes; some of it was treated with sewage effluent and sludge, and some was not. The scientists then tried seeding various plants in both types of boxes. Weeds wouldn't even grow in the unsludged spoils, but the treated material supported trees, grass, and legumes. The experiments confirmed that Pennsylvania could also benefit from reclamation with sewage wastes, possibly by pipeline from the big population centers of the East.

The idea that sewage sludge, if widely accepted by farmers, could become the nation's main fertilizer can stir resistance among commercial fertilizer people. But they only have to look at the statistics to stop worrying. If all municipal sewage in the United States were given secondary treatment, all the sludge produced would help only two tenths of one percent of the nation's 465 million acres of croplands. In fact, probably most of the sludge would be used to upgrade poor land, like the mining areas, where commercial fertilizers could never be sold.

20

A Pilot Project
for the Nation

Some six thousand acres of poor sandy soil a few miles
from the eastern shore of Lake Michigan have become
the test site for the living filter on a grand municipal
scale. It will be observed, analyzed, acclaimed, and
perhaps denounced for many years. Regardless of the
outcome, it's a brave, pioneering effort to put nature's
oldest principles to work with new techniques in a
big way.

The project is at Muskegon County, Michigan,
where community leaders had the courage in the
1960's to risk a sharp turn off the avenue of conventional sewage treatment that fails to arrive at its goal
of water pollution control. In 1973 the thirteen communities of Muskegon County took all their raw
sewage to nature for treatment in mechanically aerated

lagoons followed by storage lagoons, a system of disinfection and finally land treatment, using the living filter as the equivalent of the long-promised but seldom realized tertiary treatment by mechanical-chemical techniques.

The pioneering venture comes at the end of more than a century of ecological degradation on the shore of Lake Michigan. In the middle of the nineteenth century a thriving lumbering industry was established in the area, which was rich in pine forests that flourished on the sandy soil. But this boon was over by the turn of the century, because clear cutting had destroyed the forests.

Muskegon County was left with foundries that had supplied castings for the sawmills, and they simply found new business in the burgeoning industry of the area. The foundries were well situated because good sand for making forms was plentiful around Muskegon. But sand mining and industrial waste disposal further scarred the countryside, and foundry smoke dirtied the air.

Regardless, Muskegon prospered on and off for the first half of this century. A fortuitous oil discovery in 1927 helped the area through the Great Depression, and World War II kept the foundry business going. But then troubles piled up for the industrialized area. The oil was gone a dozen years after discovery, and the old foundries couldn't compete with modern ones in other areas. The county, with Lake Michigan on one side and several once-beautiful lakes in the middle, could have been one of Michigan's most lovely areas, except for desecration of the land, air, and water over a century of exploitation.

By the 1960's the adverse impact was evident. The unemployment rate reached twice the national average. Tourism that might have helped was blocked by the deterioration of the area's natural attractions. Agriculture had never been developed. People who could have worked to restore the county were inclined to leave. It was evident that salvation could come only from a full-scale campaign against the environmental degradation.

New leaders in the area formed the Muskegon County Metropolitan Planning Commission, with Roderick Dittmer as director of planning. The commission recognized immediately that a key to reversing the county's economic descent was environmental restoration. Central to this problem was water pollution, which required dealing fast and effectively with the waste water from 157,000 people in thirteen large and small municipalities, including Muskegon itself. Waste water purification was also urgent because of pressures from an official pact between Wisconsin, Illinois, Indiana, and Michigan, directed at keeping Lake Michigan clean by reducing the inflow of sewage phosphates 80 percent by 1972.

Dittmer and his colleagues soon concluded that to solve the county's waste water problems fully and finally required more than the usual, piecemeal approach which the conventional sanitation engineer was almost certain to recommend. In search of a more innovative program, Dittmer went to visit a notably innovative man at the Center for Urban Studies at the University of Chicago, John R. Sheaffer. He was a researcher and urban planner who had become well-known for the design of Mt. Trashmore, an artificial mountain of solid waste covered with native clay in DuPage County, Illinois, which would one day become a ski area.

Sheaffer, who was noted for his enthusiasm, saw the Muskegon problem as a challenge that might be met with some ideas that had come to him from two sources.

While a graduate student at the University of Chicago, he had heard Warren Thornthwaite discuss his Seabrook Farms land treatment design, and the student had wondered why the idea had not been adapted on a large municipal scale.

Subsequently Sheaffer began learning firsthand about the developing program at Penn State from a friend and one of its leaders, R. R. Parizek. The project, Sheaffer realized, was pointing the way to the municipal application that Thornthwaite had prompted him to think about.

An investigation at Muskegon convinced Sheaffer that a comprehensive waste water program based on the living filter could offer a three-fold return. First, it could solve the waste water problem with finality. This in turn could help open up the recreational potentials in three polluted lakes: Muskegon, Mona, and White. And lastly, the waste water could improve the county's poor soil and create an agricultural potential for the area.

Working with Dittmer and the Planning Commission under a forceful chairman, Michael Kobza, and a forward-looking County Commission chaired by Charles Raap, a program was outlined for ending water pollution in the county. Meanwhile Sheaffer introduced one of the most progressive engineers of the area, William J. Bauer of Chicago, and the county retained his firm's services to develop a waste water treatment system that could produce the benefits that the leaders now foresaw.

Sheaffer, who incidentally had helped organize SEMCO, the firm transporting sludge from Chicago to the farmland in Arcola, Illinois, was already acquainted with Muskegon County. Pursuing his interest in disposing of Chicago sludge, he had once thought the county might utilize sludge barged across Lake Michigan to upgrade its sandy soil in hopes of developing agriculture. Now, however, Sheaffer found himself instead with the objective of a full-scale system using the county's own sewage wastes to accomplish the same purpose, plus developing the added benefits seen by Sheaffer.

The plan set forth by Sheaffer called for bringing all the county's raw sewage together and pumping it eastward away from Lake Michigan and the three inland lakes to a site some fifteen miles distant, a sandy, almost barren area with few inhabitants. Here he proposed that the county establish a huge land treatment area to provide the gamut of full sewage treatment, from primary to tertiary, turning the raw waste water into drinking water while growing agricultural crops that might return money to the county as well as offer raw materials for new industry. It would start with lagoons and end with the living filter, a total package worked out with nature. Bauer was given the job of actually designing it.

But the design problems were perhaps the simplest

of all for the Muskegon project. In fact, Bauer argued personally that the whole affair was nothing more than a bag of old, well-proven tricks. Getting all the sewage to the site was only a matter of using standard pumps and pipes. Lagoons, he said, were nothing new and just because the Muskegon versions would be exceptionally large didn't mean they wouldn't work. And the living filter? Why that was just farming, said Bauer, using conventional irrigation techniques and nature's age-old capability for water purification. Here again he denied that bigness was bad. If a few square feet of soil would clean water, he stated, so would thousands of acres, if used intelligently.

As reasonable as it seemed to the proponents, the Sheaffer plan was altogether a radical departure from sewage treatment as the American mind had come to accept it. Changing that mental set in the right minds was probably the biggest accomplishment of all.

There was immediate opposition to the plan in many quarters of the county. Some of the towns had already been convinced by sanitation engineering consultants that conventional secondary treatment was the only safe, proven way to proceed. It was particularly difficult moving those town officials toward the living filter. Industries that contributed a large share of the county sewage, and would thus be paying a big portion of the treatment costs, didn't rush to acclaim the project. And the Michigan State Water Resources Commission, comfortable and committed to conventional methods, was not at all pleased with the proposals at Muskegon County.

Slowly, a step at a time, the opposition was softened

or reversed with the force of facts and reasoning. At one point, Parizek was invited by Sheaffer to come to Muskegon with a film and an explanation of the Penn State project. This prompted a group of officials to take their doubting colleagues to State College for a firsthand, personal look, and many came back as strong advocates of land treatment.

Sheaffer felt that a major turning point was reached when the corporate director of air and water pollution for one of the county's largest industries concentrated on the new idea. He was N. J. Lardieri of the Scott Paper Company which owned the S. D. Warren Paper Company in Muskegon, and his assessment of the proposal was positive. Lardieri recognized it as a way of settling the water pollution problem for good. His reaction was echoed by other local industry leaders, and the Muskegon plan was unleashed.

Momentum toward wider approval then started to build. The project excited Michigan Congressman Guy Vander Jagt, ranking Republican member of the Conservation and Natural Resources Subcommittee of the U.S. House of Representatives. Vander Jagt was convinced by Sheaffer and the Muskegon proponents that their idea could really win the long, elusive battle for water pollution control. Before long the Congressman even had President Nixon expressing enthusiasm for the project (in a letter to Sheaffer). Soon Governor William Milliken of Michigan paid a visit to Muskegon, which was followed by state approval of Sheaffer's plan.

Such action brought more than official approval. It

attracted a lot of money, including a research and demonstration grant of more than $2 million from the U.S. Environmental Protection Agency, the largest single grant ever awarded by the agency. The total cost of the construction was estimated at $36 million. Nearly half the amount was to be raised locally by the sale of bonds and the remainder was to come from state and federal funds.

But then it turned out that official acceptance backed by a lot of money didn't add up to total public acceptance. Despite publicity and public meetings to explain the new idea, the old distasteful notions about sewage persisted, especially among property owners. For a long time local citizens had shown no qualms about dumping sewage in the big lakes around them, but now they were incensed about irrigating county land with waste water.

While industry and government officials were increasingly inclined to accept the proposal, a citizens' group was formed to oppose Sheaffer's plan. They let it be known that at a decisive moment they would file a legal suit to stop the project. But the county officials, in an unusual action, turned the tables and brought the citizens' group to court to prevent them from damaging the project by untimely interference with bond sales, construction bids, and state and federal grants.

The legalities became even more complicated, and there was a trial before all three of Muskegon County's circuit judges. The testimony literally served as a comprehensive assessment of sewage treatment methods, with land treatment receiving a share of attention seldom accorded it previously. The court even held an extraordinary session at Penn State, where the judges and others observed the living filter at work and heard more testimony from the university experts.

Back in Michigan the judges soon ruled unanimously in favor of the county. The evidence, they said, did not show that the proposed system would be a public nuisance or hazard to the public health. The Muskegon project was clear to proceed with construction, which began in 1971.

When completed and in full operation around the middle of the 1970's, a collection system consisting of a network of pipes and pumping stations will lead the county's raw sewage to a large concrete main five and a half feet in diameter extending eleven miles to the treatment site. There the waste water will go to three lagoons, called "aerated biological treatment cells," each fifteen feet deep and covering eight surface acres. For three days any given amount of sewage will be aerated and mixed by twelve floating churns and a half-dozen other mechanical devices in each lagoon. The effluent will then be comparable to that from a well-designed and operated conventional secondary treatment plant.

It will next flow to two extremely large storage basins, each covering 850 acres, but only 9 feet deep. The basins will hold up to 5.1 billion gallons of effluent, and they will collect it when irrigation is not possible, in the winter or during rainy spells. Bauer designed the basins to fill up from November through March, when there will be no irrigation; and to drain down from April through October, as irrigation proceeds.

Sludge will also collect as the sewage solids settle

in the storage basins. When necessary, dredges will remove the material from the bottom and pump it to the irrigation areas for soil conditioning.

Before it is used for irrigation, however, the effluent will be disinfected with chlorine and channelled to two pumping stations which will force the liquid through a network of pipes fanning out to over fifty huge irrigation machines.

Each machine will consist of a long horizontal arm, pivoted at one end like the hand of a clock. It will revolve very slowly several feet above the ground, riding on wheeled rigs something like an aircraft landing gear. The effluent will be forced through the arm under low pressure, and it will be applied to the soil and crops from along the arm by sprays directed downward to minimize the aerosol effect. The radii of the machines will range from 750 to 1,300 feet, and their speed of rotation will be adjustable so it can take from one to seven days for the sprays to cover a full circle.

Bauer's design allows for treating up to 43.4 million gallons of sewage per day by 1992, when the projected population for the county will be 170,000. The eventual application rate for effluent (plus rain) is to be about four inches per week whenever irrigation is possible.

After infiltration of a few feet of soil, some of the purified effluent will remain with the area's groundwater, but the surplus will be caught by a large underground network of perforated drain pipes which will lead the water to the area's natural streams and on to Lake Michigan. If Bauer's design works as expected, the drainage water could even serve as a domestic supply because it will meet drinking water standards set by the U.S. Public Health Service.

In the first five years of operation an extensive agricultural research project will be conducted on the large site to determine the best crops for land treatment and community benefits. Sheaffer estimates that the agricultural profits could bring an annual return of $360,000 or more to the county.

He also envisioned part of the huge area as a possible site for a nuclear generating plant which could use the waste water for cooling purposes before it was applied to the land for irrigation. This could solve a major difficulty of nuclear generation, thermal pollution (caused when used cooling water is discharged into a stream or lake and endangers aquatic life by raising the temperature). The thermal effect would make little or no difference to the spray irrigation process. Sheaffer visualized the area as earning as much as $2.5 million a year in cooling charges and location fees from such a power plant, and perhaps other heavy industry in need of such water.

A third source of income, considered by Sheaffer and eventually approved for inclusion in the site, would come from solid waste disposal on the large land treatment site. As with Mt. Trashmore, he foresaw the building of hills, a "sculptured landscape," from solid wastes transported to the site and arranged so as not to interfere substantially with the irrigation. On the basis of a $2-per-ton disposal fee he estimated that this service could earn $300,000 annually.

Of course, the actual income from the Muskegon project will not be known until several years after the operation begins. Hopefully, it will help defray a large share of the county's sewage treatment costs. But even when the bookkeeping is done, the profits and losses in dollars will not necessarily reflect all the benefits that the project will bring to Muskegon County, from new jobs to environmental restoration.

While county officials deeply committed to their new sewage treatment system may not care to have it presented as a pilot project, it certainly is that from a national viewpoint. It could change sewage treatment forever. In recognition of the innovative quality of the Muskegon project, it was hailed by the National Society of Professional Engineers as one of the ten outstanding engineering achievements of 1972. Interestingly, land treatment projects took two of the ten places, with Chicago's Prairie Plan (the use of sludge for land reclamation) being the second.

Such change, of course, is controversial. The Muskegon project faced that reality before a shovel of dirt was moved. It was a lively topic among sanitation authorities, including land treatment proponents, across the country. Essentially the opponents seemed to be saying that it was too big—why, they weren't sure. Meanwhile Sheaffer, Bauer, and others directly behind the project remained supremely confident because they and nature were working together. The square footage made little difference. Nature knew how to work on the grand scale.

PART 7

An Epilogue
on the Future

21

The Public Demand
for an Alternative

Not long ago *The American Biology Teacher* published an article, "My Town, My Creek, My Sewage." The author, John H. Woodburn, a high school science department chairman in Rockville, Maryland, chided biology teachers for neglecting the subject of sewage, which is really pertinent to their curriculums. "Obviously, sewage and how to get rid of it hasn't been a favorite topic of study in many circles," Woodburn wrote. "The topic has long been hidden by queasiness — the best of all ways to delay the solution to a social problem."

He continued. "The stigma attached to sewage shows up in many ways. In biology textbooks, children are taken to the faraway Arctic tundra, tropical rain forests, or the pampas of the Argentine to appreciate the interdependence of species in ecosystems; but the same textbooks may ignore the supremely elegant examples of ecology in a smoothly operating sewage-treatment system."

The failure of people to deal with the subject — for lack of education, because of psychological hangups, or for various social reasons — has left it up to "experts" who function in tight, "sanitary" compartments. Relatively untouched by the leavening of public discussion and criticism, the experts, from the fraternity of sanitary engineers to state, local, and federal water pollution control officials, have often suffered from myopia. It is evident in several ways.

• They have usually been guided by a single purpose in sewage management: dispose of it in the most expeditious manner that doesn't offend the public nostrils or endanger public health.

• They have been so intently absorbed in this purpose that they seem to have missed how human waste is inevitably tied to a broad range of ecological needs and concerns.

• Their sense of geography and the fabric of earth and water has been local, as if the effects of human wastes stop at town, county, and state lines.

• Most important, their myopia has kept the compartment dwellers from exploring alternatives to the conventional sewage treatment systems that are beset by operational problems and failure to control water pollution.

Meanwhile, nonexperts outside the sanitation field, tired of the insiders' failures, have frequently pressed forward on their own to solve serious water pollution problems. In fact, some of the most notable accomplishments in the sanitation field in recent times have been prompted by the outsiders. The renowned Santee project, described in an earlier chapter, is a prime example, where Ray Stoyer, a former appliance dealer, provided the leadership. Even the technological accomplishments at Lake Tahoe were forced into being by citizens who absolutely demanded that their beautiful lake be saved from sewage, and then the key person in the design came not from the sanitation field but from the water treatment field.

There are numerous other less publicized examples. In Boulder, Colorado, for instance, a subdivision on Left Hand Creek treats its sewage with a spray irrigation system that saved the stream from pollution. It probably wouldn't be there except for a downstream mother, Sandy Cooper, who was determined that the development's secondary sewage effluent was not going to pollute the creek where her children played. She organized fellow citizens who studied the possibilities, settled on land treatment, and insisted that the housing development make use of it. After public hearings, petitions, letters to the editors, and other forms of public pressure, the citizens had their way, and Left Hand Creek remains unpolluted.

But in the great majority of cases where such citizen action does not intrude, the "experts'" limitations continue to wield influence. When domestic and industrial wastes become a problem, a solution is usually sought by officials whose area of concern is confined to a single town, city, or metropolitan area. From the beginning they move toward a standard, predictable solution: get rid of the sewage, out of sight and out of mind, at the least possible cost. For expert assistance they retain a firm of consulting sanitary engineers who determine the volume and kind of sewage involved and then offer one alternative, a conventional primary-secondary activated sludge plant. If at all possible it will be built beside the nearest river on a site to which the sewage can flow by gravity, and where the current will sweep the partially purified effluent downstream—perhaps into the next town where the same "local" problem is being considered in the same way.

The single-purpose approach with its single solution has failed, and the public that suffers the effects of water pollution is—or should be—pounding on the compartment doors for other solutions. Some of the people are, and more should be, calling for land treatment. As we have seen, the living filter offers an effective multi-purpose alternative that is seldom, if ever, considered by the experts in whose hands we have left the battle of water pollution control.

A most emphatic push to come up with alternatives to conventional methods appeared with passage of the Federal Water Pollution Control Act Amendments of 1972. The law, which evolved from a long, intensive series of Congressional hearings on water pollution control, established a national goal of eliminating the discharge of pollutants by 1985. The administrator of the act is the Environmental Protection Agency (EPA), and the agency is directed by the law itself to seek alternatives to the conventional means of treating sewage. This is especially evident where the legislation establishes an important demonstration program for the Great Lakes.

"This program," the act states, "shall set forth alternatives for managing waste water on a regional basis and shall provide local and state governments with a range of choice as to the type of system to be used for the treatment of waste water. These alternative systems shall include both advanced waste treatment technology and land disposal systems including aerated [lagoon] treatment-spray irrigation technology and will also include provisions for the disposal of solid wastes, including sludge. . . ."

As the federal law establishes funds for construction, it again directs the administrator to seek alternatives, especially systems that might produce revenue from recycling methods involving agriculture and forestry. To the acclaim of those who advocate land treatment, the act states: "The Administrator shall encourage waste treatment management which combines 'open space' and recreational considerations with such management."

Incidentally, the fact that the new federal act so strongly advocated land treatment was the result of an amendment successfully offered from the floor of the House of Representatives by Congressman Guy Vander Jagt whose home district includes Muskegon County. The amendment was offered just prior to the House vote on the clean water bill and was retained by the House-Senate committee of conference that decided on the final version of the legislation.

The serious intent of the law's advocacy of land treatment is clear as it proclaims that after June 30,

1974, federal funds — which represent many billions of dollars — will not be granted to state and local sources for "the erection, building, acquisition, alteration, remodeling, improvement, or extension of treatment works" unless the applicant has satisfactorily demonstrated to the EPA that all alternatives have been "studied and evaluated" and that reclaiming and recycling of water has been taken into account.

While the admonition to explore alternatives is now law, it is far from resulting in a *fait accompli*. In fact, the search for alternatives may be no more than an empty gesture if the exploration is not demanded by an informed public. Resistance to change in the water pollution control field is substantial, from the EPA in Washington down to the smallest town's sewer commissioner. It was described as follows by David Zwick, director of the Nader-sponsored Fisherman's Clean Water Action Project: "The pollution control 'experts' on whom every level of government is accustomed to relying to solve their water pollution control problems, are, with few exceptions, heavily committed — as a result of their training, life-long career work, and often a vested financial interest — to conventional primary, secondary, and tertiary (advanced) wastewater treatment."

The limitations of their timeworn expertise were discussed when Eugene T. Jensen, then the Chief of the Water Quality Office of the EPA, addressed the American Society of Civil Engineers National Specialty Conference in Los Angeles. Jensen said he was ashamed to admit it, but the "old pros" of water pollution control were lagging. They had been willing to settle for too little, to tolerate poor operations and to look for problems instead of solutions when called up to find methods of reusing waste water. The result, said Jensen, "is that the cases in which a major [polluted] stream or lake has actually been restored can be counted on one hand." He went on to conclude that, "We, the professionals in the field of water pollution control, are going to have to change ourselves, our concepts, and our way of doing things."

While the Water Pollution Act called for such change, it was already being stimulated on the national level by a surprising program initiated by the Army Corps of Engineers in late 1970. It was a pilot program "designed to develop and evaluate the full range of alternative systems available to alleviate wastewater problems." The program, approved by the Federal Office of Management and Budget and the Public Works and Appropriations Committees of Congress, was to be particularly concerned with exploring two alternatives: the conventional methods and land treatment. A thesis of the new project was explained by Thaddeus R. Beal, Under Secretary of the Army, as follows:

"The pilot wastewater management program recognizes that the environment must be treated as a total system with air, land, and water interacting and affecting and being affected by man. For planning purposes, pollutants are treated simply as resources out of place. For example, the nutrients in wastewater which accelerate eutrophication of our nation's water bodies became valuable fertilizer when applied to the land."

To dedicated conservationists such words were

pure harmony, but they had to be suspect coming from the Corps of Engineers, whose public works on America's rivers and harbors had been blamed for some of the most serious ecological damage in the nation's history. Only recently a proposed study of water resources by the corps had been described as "approximately analogous to asking Jack the Ripper to make a study of prostitution." Now, with surprisingly enlightened ideas, the organization was taking what sounded like an intelligent look at the nation's immense waste water problems.

When a few reporters looked into the new pilot program, they found John R. Sheaffer, the catalytic force who had started the Muskegon land treatment project. Remaining a civilian, he had come to the office of the Secretary of the Army from the University of Chicago through an annual visiting professorship that the Army had previously established. But he also arrived with a title (Scientific Adviser) that established him at a rank equivalent to a two-star general. This was important for getting things done in the rank-conscious Army.

Sheaffer was certainly aware of the corps' blackened image among conservationists, but instead of merely deploring the organization, he chose to see its strengths rather than its failings. They were considerable, and they could help accomplish what he felt was needed in waste water management, a broad, serious look at land treatment as an alternative to the old-line methods that were proven failures. His arrival in Washington in September 1970 was soon followed by initiation of the pilot program.

It began by using the corps' vast array of talents to conduct waste water management studies in five major urban regions of the United States: the Merrimack River Basin extending north of Boston, the Cleveland-Akron area, southeast Michigan around Detroit, the Chicago region including northwest Indiana, and a California region encompassing San Francisco, Sacramento, and Stockton.

Not long after the program had begun, the corps at the request of the White House issued a special preliminary report presenting three alternative waste water programs for the Chicago metropolitan region and giving comparative costs. Two would follow the conventional, technological approach of developing advanced waste water treatment plants designed simply to clean sewage enough so that it could be dumped without polluting natural waters. The third alternative was a daring proposal *à la* Muskegon.

It called for transporting all raw sewage from the Chicago region's municipal and industrial sources to a 448,000-acre irrigation site on poor farmland southeast of the city, partly in Illinois but mostly in Indiana. It would flow there by gravity through deep tunnels. At the site the sewage would be pumped into vast lagoons for treatment and storage. In the growing season the effluent would be spray-irrigated on some 371,000 acres of cropland. Sludge dredged from the lagoons would also go to the soil. The water cleaned by the plants and the land would be collected by underdrains and returned to the Chicago area to supplement its surface and groundwater supplies.

A cost analysis of the three alternatives indicated

that the land treatment system would be much less expensive than either of the advanced waste water treatment alternatives. The land treatment site might also offer the other benefits considered in Muskegon. It might be the home of a nuclear power plant and heavy industry requiring a lot of cooling water. And it had other possible advantages, ranging from revenue-producing crops to recreation areas, solid waste disposal, and ecological compatibility with nature.

The preliminary report revealed that, right or wrong, feasible or unfeasible, someone was finally questioning the old, one-purpose, local approach which had left the Chicago metropolitan area with nearly 350 sewage treatment plants that still didn't add up to adequate water pollution control. Now people could see an alternative to the conventional, piecemeal management, and this in itself could be a key step toward a comprehensive, complete solution to Chicago's old waste water problem.

The corps' program was enlarged to study waste water management in more than a dozen and a half smaller regions. While the army organization was not authorized to build treatment systems, its exploration of alternatives was certain to force people, from professional sanitarians to concerned laymen, to recognize that nature may often have the workable alternative to the "old pros'" less than satisfactory approach to water pollution control.

While the corps may demonstrate the wide possibilities of land treatment in certain specific areas, the actual development of such systems must be worked out at the community level, as was true in Muskegon

County. For this to happen the citizens who suffer most from water pollution and pay dearly for its control must become well enough informed to demand the exploration and development of alternatives called for by the federal act of 1972. People who want clean water and the benefits of green growth that could accompany land treatment must insist that public sanitation officials seriously assess the possibilities.

To date, this depends largely upon the consulting engineers who ordinarily provide the officials with expertise on sewage treatment. Unfortunately, most are reluctant even to consider land treatment as a viable alternative. Their reluctance is explained by at least three reasons.

First, they see the idea as "land disposal" which the engineering profession discounted years ago as an unworkable, unsafe practice of the past. At best sanitary engineering textbooks may mention land disposal in a brief historical note before they go into extensive technical detail on conventional treatment plants. So from the start an engineer is led to think that taking sewage to the land is an archaic method of dumping waste water.

Second, the conventional consulting engineer, should he even grant the validity of land treatment, is still likely to dismiss the idea by saying that we don't have the land to do the job. This excuse is based on several false assumptions discussed or alluded to earlier. One of the main assumptions is that we are unalterably confined to treating waste waters within the political boundaries of where the sewage is generated, thus in urban areas producing the greatest volume of sewage, potential land for treatment is in short supply and prohibitively expensive. True—but we are no longer bound by such political confines. In fact, the barriers are to be removed by the federal water pollution control law of 1972 which is designed to force states to manage waste water on an expansive regional basis. In this framework, nutrient-rich waste water can flow from urban areas back to the rural lands from where the nutrients (via agricultural products) came in the first place (the Chicago Prairie Plan and Muskegon project, for example). When we seriously consider taking an otherwise wasted resource back to the country for beneficial uses, the no-land argument disintegrates. There land is available and the resource has economic value.

The fact of the matter is, we haven't even started to explore the possibilities of land-use planning as it relates to waste water management. Why can't public sanitation officials work with private farmers (as they have in Paris, France, and Lubbock, Texas, for decades) to raise beneficial crops, from hay to trees, while purifying the waste waters of towns and cities? They can if they try. Why can't the badly needed development and preservation of green open space in and around our urban areas be coordinated with the treatment of the area's waste waters with nature's help? It can if we try. Around our vast interstate highway system and other land-using systems for transportation, electric power transmission, and pipeline rights of way we have millions and millions of square miles of land that might be used, beautified, and put to productive use with the nutrients of waste water. Certainly, with the application of thought and imagination, land can be found for treating our waste waters with beneficial results—as illustrated by a California research project which is demonstrating that waste water may be valuable for keeping firebelts green in dry areas.

One of the most imaginative projects that helps to demolish the no-land argument is at Walt Disney World in Florida. The entrepreneurs, in cooperation with the University of Florida, are developing large pieces of land around their recreational properties to treat the waste waters of Disney World itself. They will raise crops of various kinds, grasses, ornamental plants, and many varieties of trees. Animals from the complex will graze on parts of these lands, and viewing roads will allow visitors to observe the area, which is expected to become one of the most beautiful, exotic displays of green growth in the United States.

Lack of land is really a weak argument for opposing land treatment. Acreage is available at costs that can be afforded and with potential benefits that make the use truly worthwhile. Actually there's a need to break open the barriers of false assumptions and antiquated ideas that keep us from exploring the possibilities.

A third factor causing many engineers to favor conventional waste water treatment over land treatment is that the latter is much more demanding of time and talent than the former. The design of an effective land treatment system requires the application of numerous skills, many of which are not needed in planning and building conventional systems. For example, soil specialists, agronomists, geologists, agricultural technicians, and other disciplines outside the ordinary sanitation engineer's competence may be required even to select an effective site. Most consultants aren't prepared to provide such services, nor are they comfortable with the idea of using them.

Finally, the consultants' fee structure (usually a percentage of the capital cost of a project) also becomes less certain in developing land treatment systems. Conventional treatment plants have a sameness from installation to installation, so engineers can safely predict their time expenditures and income. Land treatment, which depends on a whole range of variables from groundwater hydrology to climate, naturally lacks this sameness. The engineers' investment of time is not so easily predicted, and in turn, his financial gain is not as certain.

These and other factors mean that engineering firms willing to tackle land treatment installations are scarce. The Bauer Engineering Company, designer of the Muskegon project, is, of course, one of the scarce examples. The organization continues applying its multiplicity of skills to land treatment projects. After Muskegon, for example, Bauer, working with a Boston firm, began planning a Cape Cod system with problems far different from the ones found in Michigan. They were hired by the coastal town of Falmouth, Massachusetts, where citizens had organized to save beautiful Vineyard Sound from pollution by sewage.

A Denver firm, Wright-McLaughlin Engineers, is becoming widely known for its interest and competence in land treatment, with several projects to their credit around Colorado. They state that: "The land treatment of sewage effluent is environmentally sound, it has the capability of capturing the imagination and support of the average citizen and it meets better than any other waste-water treatment method the challenges of basic environmental planning principles."

Still another example of a firm prepared to design and develop land treatment systems is Williams and Works of Grand Rapids, Michigan. By the early 1970's the company could point to its land treatment projects in over a dozen Michigan communities. In general, raw sewage was treated in lagoons and the effluent was applied to the soil, to irrigate various crops and to be purified. Both flood and spray irrigation techniques were used. Some of the expertise was drawn from the scientific developments at Penn State.

Williams and Works' capability for land treatment is attributable to the firm's having an unusual variety of skills required by such projects. Their experts include geologists, geochemists, community planners, authorities on government funding and regulations, soil specialists, drainage experts, authorities on lagoons and irrigation, and many others from a total staff of 160. Williams and Works' staff even includes talent to help clients with the key problem of gaining public acceptance for land treatment projects. Town officials are advised on how to conduct public education programs, from the running of public hearings to the use of the media, to build the community confidence necessary for what might seem like a major, untried departure in sewage treatment.

If the supply of such services in the nation is to

grow, the demand must increase from an informed public that desires to have the alternatives of land treatment thoroughly explored and developed wherever possible. If demand for this kind of comprehensive waste water management persists, experts capable of using the living filter are certain to develop at an increasing rate. Textbook authors will have to cover the alternative of land treatment, and engineering courses will have to offer training in the necessary skills. And finally public officials at many levels will begin receiving expert advice that considers human waste in the total context of nature's basic cycles of life, rather than in the confined view of its being a burdensome nuisance to be rid of.

But as John Woodburn, the high school science teacher, warns, the best way to delay settling our difficult water pollution problem is to continue avoiding the subject of sewage—to go on hiding it "by queasiness." If sewage is to become a blessing instead of a blight, the subject of human waste can no longer be neglected in schools, the media, service clubs, politics—wherever people are concerned about what we have done to our once beautiful rivers, streams, and lakes, and to ourselves in the process.

Appendix A

Number of confirmed land treatment facilities by state as of 1971 (based on survey by Center for Study of Federalism, Temple University)

	SOURCES OF WASTE WATER		
	Municipal	Industrial	Agricultural
Alabama (B)			
Alaska (B)			
Arizona (A)	19		many
Arkansas (D)		3	
California (A)	260	3	
Colorado (A)	6		
Connecticut (E)	3		
Delaware (E)	1	2	
Florida (D)	3		
Georgia (E)			
Hawaii (E)			
Idaho (A)	2		1
Illinois (D)		1	
Indiana (E)		4	
Iowa (C)	1	2	
Kansas (C)	2		
Kentucky (E)			
Louisiana (E)	1	1	
Maine (C)			
Maryland (A)	9	2	1
Massachusetts (B)	5		
Michigan (C)	2	4	2

	SOURCES OF WASTE WATER		
	Municipal	Industrial	Agricultural
Minnesota (E)	1	5	14
Mississippi (E)			
Missouri (E)			
Montana (A)	5		
Nebraska (D)	1	1	
Nevada (E)	12		
New Hampshire (E)	2		
New Jersey (A)	2	3	
New Mexico (A)	2		
New York (A)			
North Carolina (E)	14		
North Dakota (A)	6		
Ohio (E)		3	1
Oklahoma (A)	3		
Oregon (E)	6	3	1
Pennsylvania (E)	1	5	1
Rhode Island (E)			
South Carolina (E)		1	
South Dakota (E)			
Tennessee (E)		4	
Texas (A)	109	3	
Utah (D)	1		
Vermont (A)			
Virginia (D)		1	
Washington (D)	10	4	
West Virginia (E)	1		
Wisconsin (A)	4	1	25
Wyoming (E)	4		
Totals	498	56	46+

(A) *States favorably oriented toward land treatment*
(B) *Neutral*
(C) *Permit land treatment with restrictions*
(D) *Negative orientation*
(E) *No judgment possible from data*

Appendix B

State water pollution control agencies for information on possibilities of land treatment (compiled by Center for Study of Federalism, Temple University, 1971)

Alabama
Water Improvement Commission
State Office Building
Montgomery, Alabama   36104

Alaska
Division of Environmental Health
Department of Health and Welfare
Pouch H
Juneau, Alaska   99801

Arizona
Environmental Health Services
Department of Health
1624 West Adams Street
Phoenix, Arizona   85007

Arkansas
Arkansas Pollution Control Commission
1100 Harrington Avenue
Little Rock, Arkansas   72202

California
State Water Resources Control Board
1416 Ninth Street, Room 1140
Sacramento, California   95814

Colorado
Water Pollution Control Division
Colorado Department of Public Health
4210 East Eleventh Avenue
Denver, Colorado   80220

Connecticut
Department of Environmental Protection
State Office Building, Room 539
Hartford, Connecticut   06115

Delaware
Division of Environmental Control
Department of Natural Resources and Environmental Control
P.O. Box 916
Dover, Delaware   19901

District of Columbia
District of Columbia Department of Human Resources
1875 Connecticut Avenue, N.W.
Washington, D.C.   20009

Florida
Department of Air and Water Pollution Control
Tallahassee Bank Building, Suite 300
315 South Calhoun Street
Tallahassee, Florida   32301

Georgia
State Water Quality Control Board
47 Trinity Avenue, S.W., Room 609
Atlanta, Georgia   30334

Hawaii
Environmental Health Division
Hawaii Department of Health
Box 3378
Honolulu, Hawaii   96801

Idaho
Environmental Improvement Division
Idaho Department of Health
State House
Boise, Idaho   83707

Illinois
Illinois Environmental Protection Agency
2200 Churchill Road
Springfield, Illinois   62706

Indiana
Stream Pollution Control Board
1330 West Michigan Street
Indianapolis, Indiana   46206

Iowa
State Department of Health
Lucas State Office Building
Des Moines, Iowa   50319

Kansas
Division of Environmental Health
Kansas State Department of Health
Topeka, Kansas   66612

Kentucky
Kentucky Water Pollution Control Commission
275 East Maine Street
Frankfort, Kentucky   40601

Louisiana
Louisiana Stream Control Commission
P.O. Drawer FC, University Station
Baton Rouge, Louisiana   70803

Maine
Environmental Improvement Commission
State House
Augusta, Maine   04330

Maryland
Environmental Health Services
State Department of Health and Mental Hygiene
2305 North Charles Street
Baltimore, Maryland   21218

Massachusetts
Division of Water Pollution Control
Department of Natural Resources

Leverett Saltonstall Building
Government Center
Boston, Massachusetts   02202

Michigan
Michigan Water Resources Commission
Stevens T. Mason Building, Station A
Lansing, Michigan   48926

Minnesota
Minnesota Pollution Control Agency
717 Delaware Street, S.E.
Minneapolis, Minnesota   55440

Mississippi
Mississippi Air and Water Pollution Control Commission
P.O. Box 827
Jackson, Mississippi   39205

Missouri
Missouri Water Pollution Board
P.O. Box 154
Jefferson City, Missouri   65101

Montana
Montana Water Pollution Control Council
Division of Environmental Sanitation
Montana Department of Health
Helena, Montana   59601

Nebraska
Nebraska Environmental Control Council
State House Station   Box 94653
Lincoln, Nebraska   68509

Nevada
Department of Health, Welfare and Rehabilitation
Nye Building, 201 South Fall Street
Carson City, Nevada   89701

New Hampshire
Water Supply and Pollution Control Commission
105 Loudon Road, Prescott Park
Concord, New Hampshire   03301

New Jersey
Department of Environmental Protection
P.O. Box 1390
Trenton, New Jersey   08625

New Mexico
New Mexico Water Quality Control Commission
Health and Social Services Department
P.O. Box 2348
Santa Fe, New Mexico   87501

New York
Department of Environmental Conservation
Albany, New York   12201

North Carolina
North Carolina Board of Water and Air Resources
P.O. Box 27048
Raleigh, North Carolina   27611

North Dakota
Environmental Health and Engineering Service
North Dakota State Department of Health
Bismarck, North Dakota   58501

Ohio
Water Pollution Control Board
State Department of Health
P.O. Box 118
Columbus, Ohio   43216

Oklahoma
Environmental Health Services
State Department of Health
3400 North Eastern Avenue
Oklahoma City, Oklahoma   73105

Oregon
Department of Environmental Quality
P.O. Box 231
Portland, Oregon   97201

Pennsylvania
Bureau of Sanitary Engineering

Department of Environmental Resources
P.O. Box 2351
Harrisburg, Pennsylvania   17120

Rhode Island
Rhode Island Department of Health
335 State Office Building
Providence, Rhode Island   02903

South Carolina
South Carolina Pollution Control Authority
P.O. Box 11628
Columbia, South Carolina   29211

South Dakota
Division of Sanitary Engineering
South Dakota State Department of Health
Pierre, South Dakota   57501

Tennessee
Tennessee Water Quality Control Board
Cordell Hull Building, Room 621
Sixth Avenue North
Nashville, Tennessee   37219

Texas
Texas Water Quality Board
1108 Lavaca Street
Austin, Texas   78701

Utah
Utah Water Pollution Committee
Department of Social Services
44 Medical Drive
Salt Lake City, Utah   84113

Vermont
Department of Water Resources
Agency of Environmental Conservation
State Office Building
Montpelier, Vermont   05602

Virginia
State Water Control Board
P.O. Box 11143
Richmond, Virginia   23230

Washington
Washington Department of Ecology
P.O. Box 829
Olympia, Washington   98501

West Virginia
Division of Water Resources
Department of Natural Resources
1201 Greenbrier Street
Charleston, West Virginia

Wisconsin
Division of Environmental Protection
Wisconsin Department of Natural Resources
P.O. Box 450
Madison, Wisconsin   53701

Wyoming
Sanitary Engineering Services
Wyoming Department of Health and Social Services
Division of Health and Medical Services
State Office Building
Cheyenne, Wyoming   82001

Appendix C

List of often discussed sewage treatment systems where soil (with or without crops) is used for advanced (tertiary) wastewater treatment

Arizona
  (Following are among more than thirty municipalities using land to treat all or part of effluent.)
Phoenix (Flushing Meadows Project)
Tucson

California
  (Following are among more than 250 municipalities using land to treat all or part of effluent.)
Bakersfield
Fresno
Pomona
San Francisco (Golden Gate Park)
Santee
U.S. Marine Corp Base, Twenty-nine Palms
Whittier Narrows

Colorado
  (Following are among more than twenty municipalities using land to treat all or part of effluent.)
Aurora
Boulder (Lake of the Pines Subdivision)
Colorado Springs
U.S. Air Force Academy

Florida
Lake Buena Vista (Walt Disney World)
Tallahassee

Illinois
Arcola (sludge application)
Cook County forest preserves near 143rd and Ridgeland (sludge
    application)
Fulton County (the Prairie Plan sludge application)
Ottawa (sludge application)
Richmond (sludge application)

Indiana
E & M Ranch, Newton County (sludge application)
Terre Haute (Commercial Solvents Co.)

Maryland
Chestertown (Campbell Soup)
St. Charles

Michigan
Belding
Harbor Springs
Leoni
Muskegon
Roscommon
Wayland

New Hampshire
Lake Sunapee State Park

New Jersey
Bridgeton (Shoemaker Dairies)
Rossmore (Forsgate Sanitation)
Salem (H. J. Heinz)
Seabrook (Seabrook Farms)

Ohio
Napoleon (Campbell Soup)
Urbana (Howard Paper Mills)

Pennsylvania
Benner (Continental Trailer Court)
Chester (Green Valley Farms)

Hanover (sludge application)
Harleysville (sludge application)
Milford (Riegel Paper Co.)
St. Mary's (sludge application)
State College (Penn State Project)

Texas
　　(Following are among more than seventy-five municipalities using land to treat all or part of effluent.)
Colorado City
Diboll
Fredericksburg
Lubbock
Paris (Campbell Soup)

Appendix D

Excerpt from the "Federal Water Pollution Control Act Amendments of 1972" (Public Law 92-500, 92nd Congress, S. 2770, October 18, 1972) encouraging the use of land treatment for waste water. Drawn from pages 18–19, Section 201

"(c) To the extent practicable, waste treatment management shall be on an areawide basis and provide control or treatment of all point and nonpoint sources of pollution, including in place or accumulated pollution sources.

"(d) The Administrator shall encourage waste treatment management which results in the construction of revenue producing facilities providing for—

"(1) the recycling of potential sewage pollutants through the production of agriculture, silviculture, or aquaculture products, or any combination thereof;

"(2) the confined and contained disposal of pollutants not recycled;

"(3) the reclamation of wastewater; and

"(4) the ultimate disposal of sludge in a manner that will not result in environmental hazards.

"(e) The Administrator shall encourage waste treatment management which results in integrating facilities for sewage treatment and recycling with facilities to treat, dispose of, or utilize other industrial and municipal wastes, including but not limited to solid waste and waste heat and thermal discharges. Such integrated facilities shall be designed and operated to produce revenues in excess of capital and operation and maintenance costs and such revenues shall be used by the designated regional

249

management agency to aid in financing other environmental improvement programs.

"(f) The Administrator shall encourage waste treatment management which combines 'open space' and recreational considerations with such management.

"(g) (1) The Administrator is authorized to make grants to any State, municipality, or intermunicipal or interstate agency for the construction of publicly owned treatment works.

"(2) The Administrator shall not make grants from funds authorized for any fiscal year beginning after June 30, 1974, to any State, municipality, or intermunicipal or interstate agency for the erection, building, acquisition, alteration, remodeling, improvement, or extension of treatment works unless the grant applicant has satisfactorily demonstrated to the Administrator that—

"(A) alternative waste management techniques have been studied and evaluated and the works proposed for grant assistance will provide for the application of the best practicable waste treatment technology over the life of the works consistent with the purposes of this title; and

"(B) as appropriate, the works proposed for grant assistance will take into account and allow to the extent practicable the application of technology at a later date which will provide for the reclaiming or recycling of water or otherwise eliminate the discharge of pollutants."

Appendix E

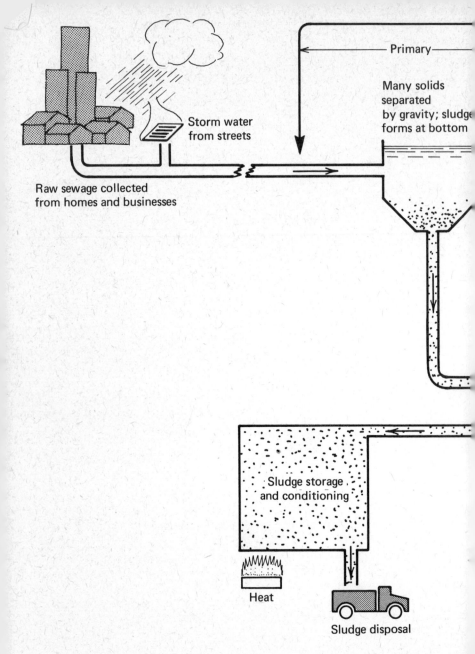

Storm water from streets

Raw sewage collected from homes and businesses

Primary

Many solids separated by gravity; sludge forms at bottom

Sludge storage and conditioning

Heat

Sludge disposal

Diagrams by Rino Dussi

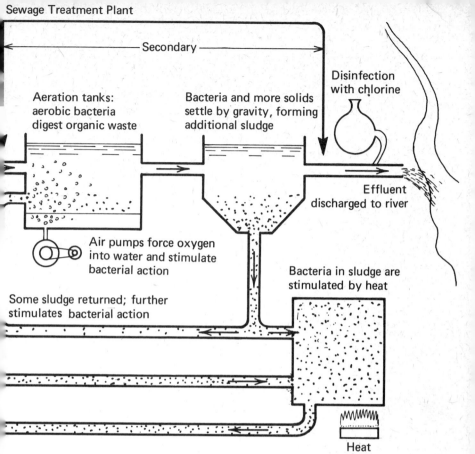

Secondary

Aeration tanks:
aerobic bacteria
digest organic waste

Bacteria and more solids
settle by gravity, forming
additional sludge

Disinfection
with chlorine

Effluent
discharged to river

Air pumps force oxygen
into water and stimulate
bacterial action

Some sludge returned; further
stimulates bacterial action

Bacteria in sludge are
stimulated by heat

Heat

1. *Conventional Primary-Secondary Sewage Treatment System*
   (Activated Sludge Process)

   - Primary treatment removes about 40 percent of pollutants in
     allowing solids to settle out by gravity.
   - Secondary treatment removes approximately 45 percent additional pollutants through bacterial action stimulated by
     aeration and returned sludge.

   This level of treatment is the highest attained or aspired to in
   most municipalities. It is the goal set for all municipalities by 1977.
   Primary-secondary treatment concentrates essentially on organic wastes, leaving a large percentage of the nutrients, phosphorus and nitrogen, that are serious causes of water pollution.

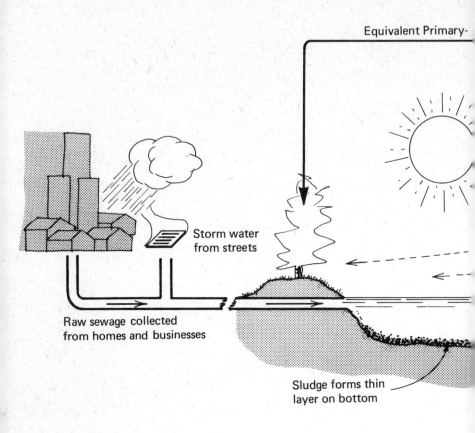

Equivalent Primary-

Storm water
from streets

Raw sewage collected
from homes and businesses

Sludge forms thin
layer on bottom

2. *Simple Lagoon System*

The lagoon is a shallow, man-made lake designed to accept a given amount of raw sewage, allowing it to pass through in a specific period of time (a month or so) to provide the equivalent of secondary treatment. A lagoon may require considerable land, but construction is relatively inexpensive and, when properly designed, operation and maintenance costs are low.

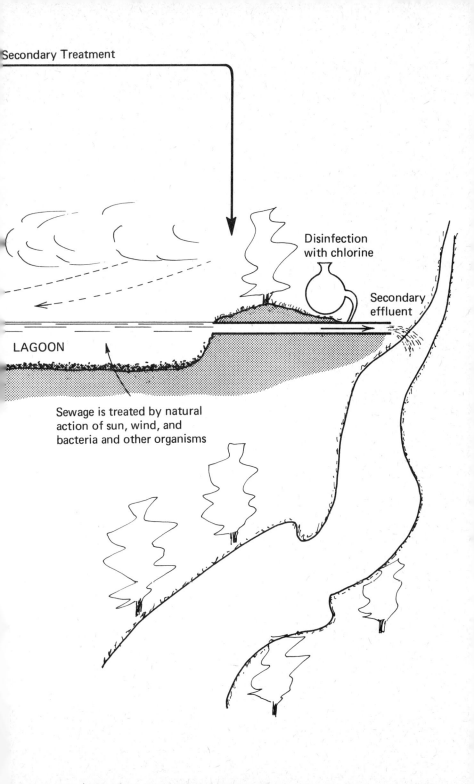

Secondary Treatment

Disinfection
with chlorine

Secondary
effluent

LAGOON

Sewage is treated by natural
action of sun, wind, and
bacteria and other organisms

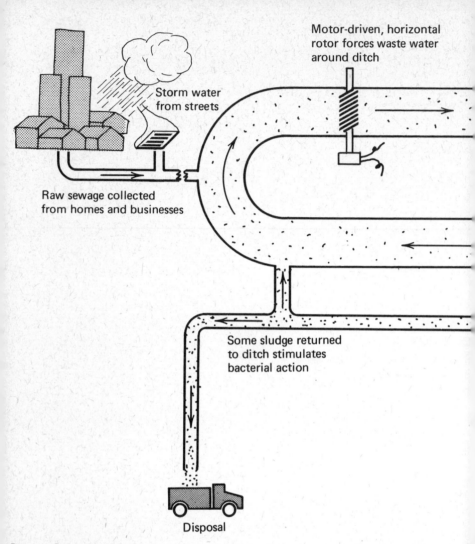

Storm water from streets

Motor-driven, horizontal rotor forces waste water around ditch

Raw sewage collected from homes and businesses

Some sludge returned to ditch stimulates bacterial action

Disposal

Pasveer Oxidation Ditch

3. *Popular European Wastewater Treatment Systems Designed in Holland*

The oxidation ditch is a biological system with bacterial action enhanced by movement of wastewater around racetrack and by re-introduction of sludge.

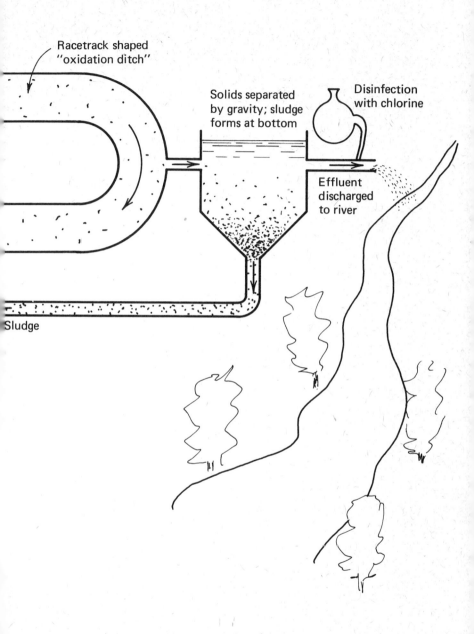

Racetrack shaped "oxidation ditch"

Solids separated by gravity; sludge forms at bottom

Disinfection with chlorine

Effluent discharged to river

Sludge

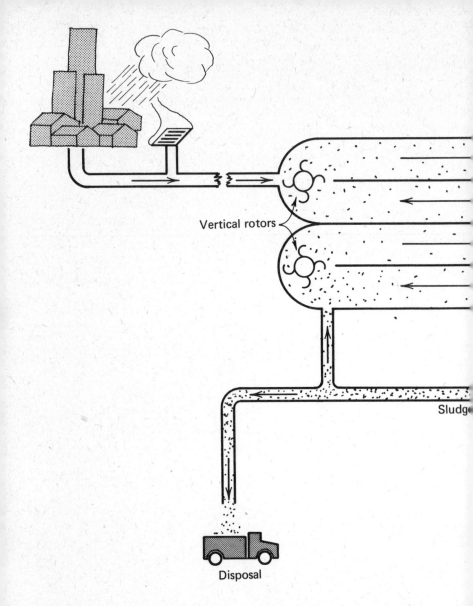

Vertical rotors

Sludge

Disposal

Carrousel

3. *(Continued)*

The carrousel is a later, improved version of the ditch from the same designers. Vertical rotors allow greater depth than in ditch so more wastewater may be treated with less land use.

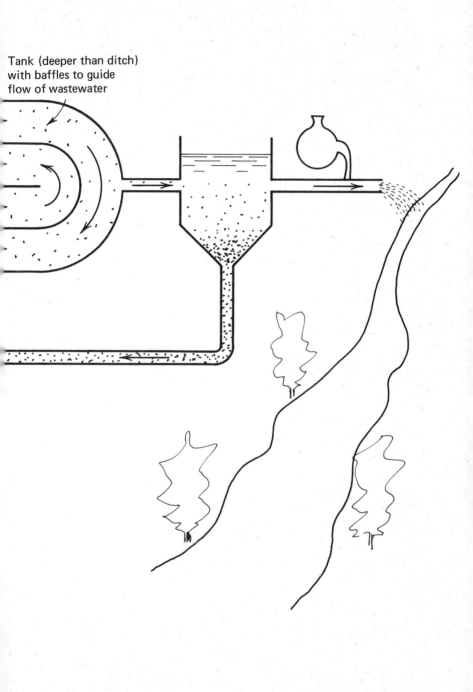

Tank (deeper than ditch)
with baffles to guide
flow of wastewater

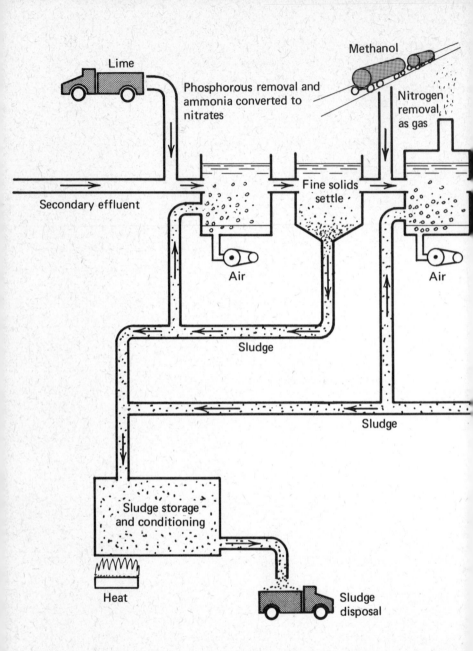

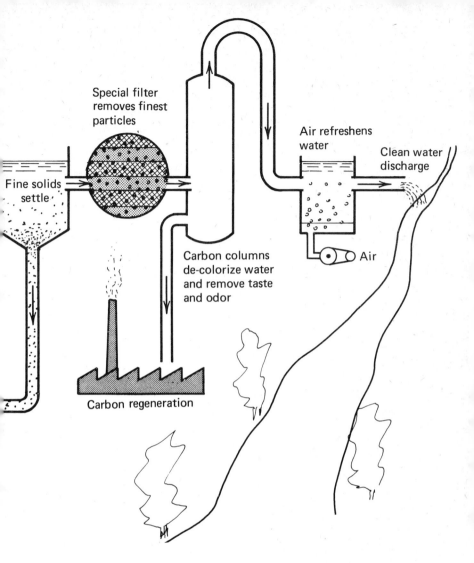

Special filter removes finest particles

Air refreshens water

Clean water discharge

Fine solids settle

Carbon columns de-colorize water and remove taste and odor

Air

Carbon regeneration

4. *Typical Proposed Advanced (Tertiary) Wastewater Treatment System*

Starting with secondary effluent, biological, physical, and chemical "unit processes" remove nutrients, taste, odor, and remaining solids down to finest particles. To reclaim one resource, i.e., water, such a system consumes many other resources—chemicals, power to operate, transportation for chemicals and sludge removal, etc. Complete maintenance from primary through tertiary systems requires many skills and large financial expenditure. And sludge still presents a waste disposal problem.

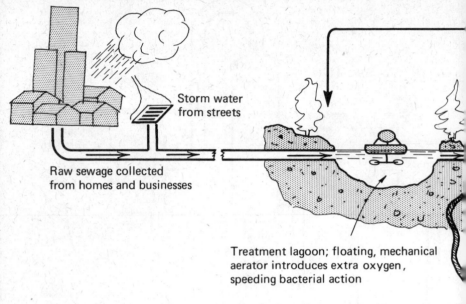

Storm water from streets

Raw sewage collected from homes and businesses

Treatment lagoon; floating, mechanical aerator introduces extra oxygen, speeding bacterial action

Advanced Treatment

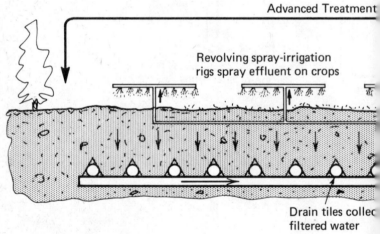

Revolving spray-irrigation rigs spray effluent on crops

Drain tiles collec filtered water

5. *Land Treatment System* (Patterned on Muskegon Plan)

Advanced treatment, following initial treatment by lagoons, depends on "living filter" with soil and crops providing complete purification.

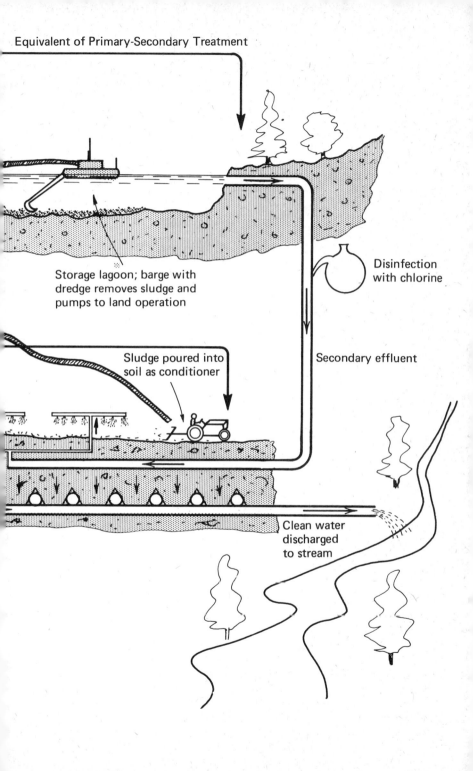

Equivalent of Primary-Secondary Treatment

Storage lagoon; barge with
dredge removes sludge and
pumps to land operation

Disinfection
with chlorine

Sludge poured into
soil as conditioner

Secondary effluent

Clean water
discharged
to stream

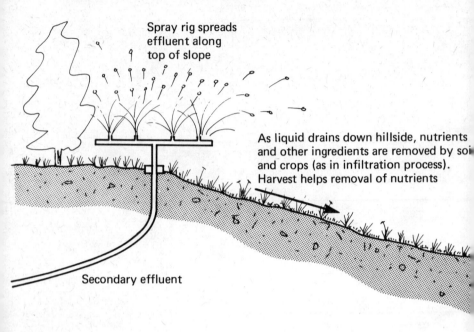

Spray rig spreads
effluent along
top of slope

As liquid drains down hillside, nutrients
and other ingredients are removed by soil
and crops (as in infiltration process).
Harvest helps removal of nutrients

Secondary effluent

6. *The Living Filter—Overland Flow*

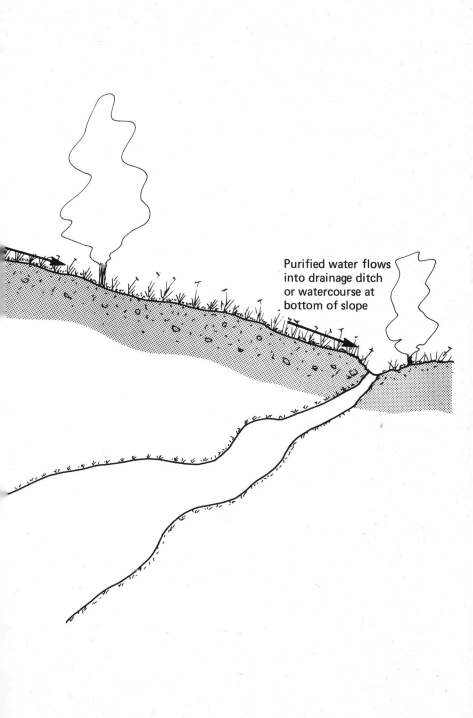

Purified water flows
into drainage ditch
or watercourse at
bottom of slope

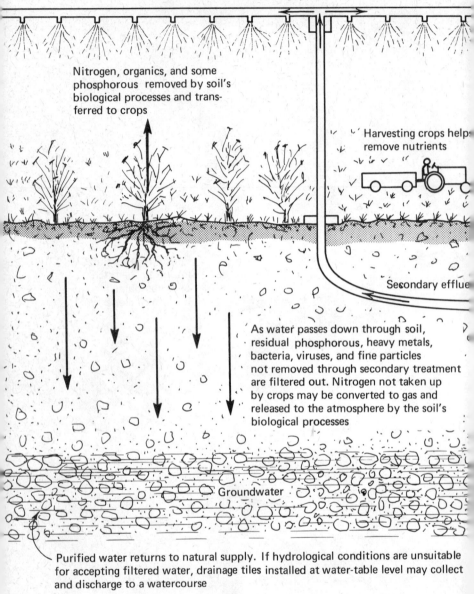

Slowly revolving spray-irrigation rig
spreads effluent on soil and crops

Nitrogen, organics, and some
phosphorous removed by soil's
biological processes and transferred to crops

Harvesting crops helps
remove nutrients

Secondary effluent

As water passes down through soil,
residual phosphorous, heavy metals,
bacteria, viruses, and fine particles
not removed through secondary treatment
are filtered out. Nitrogen not taken up
by crops may be converted to gas and
released to the atmosphere by the soil's
biological processes

Groundwater

Purified water returns to natural supply. If hydrological conditions are unsuitable
for accepting filtered water, drainage tiles installed at water-table level may collect
and discharge to a watercourse

7. *The Living Filter—Soil Infiltration*

Glossary

Activated Sludge, a secondary treatment process where primary effluent is mixed with air and recirculated sludge that is heavily laden with bacteria.

Adsorption, a process of passing pre-treated waste water through a medium such as carbon and having the dissolved organic matter concentrate on the medium surface.

Advanced waste water treatment, see "tertiary."

Aeration, mixing of air and water by physical means to increase the dissolved oxygen levels.

Aerobic bacteria, type of bacteria that requires oxygen. They are the key to the biological activity essential to properly functioning secondary treatment systems (activated sludge, trickling filters, and lagoons). They also play an important role in the action of soil in land treatment systems.

Algae, chlorophyll containing aquatic plants characterized by their ability to carry out photosynthesis and associated with taste and odor problems in water.

Anerobic bacteria, type of bacteria that grows without oxygen. They are involved in the action of septic tanks, and they are useful in the process of denitrification in soil conditions where oxygen is not present. Foul odors associated with sewers result from anerobic conditions and the digestive processes of anerobic bacteria. Such odors are prevented by avoiding exposed anerobic conditions.

Aquifer, natural underground storage area for water in stratum of permeable rock, sand, or gravel.

Bacteria, the simplest form of plant life capable of supporting all life processes for growth and reproduction—microscopic, one-celled, and colorless.

Benthic Organisms, plants or animals that live on or in the bottom materials of a lake or stream.

267

Biodegradable Organics, wastes contributed to domestic and industrial sewage that may be biologically oxidized.

BOD (biochemical oxygen demand), measure of waste water pollutional strength based on the amount of oxygen required by bacteria to oxidize organic waste. Used as an indicator of how well a sewage treatment plant is working.

Cfs, flow rate in cubic feet per second; 1 cfs = 0.65 million gallons per day.

Chlorination, the disinfection of water or waste water with chlorine.

Coagulation, the combining of solids to make them settle faster.

COD (chemical oxygen demand), measure of waste water pollutional strength based on the oxygen consumed in a chemical reaction.

Coliform bacteria, organisms found in the intestines of warm-blooded animals. Counts of these organisms, which in themselves are not disease causing, are used as a test of water that indicates the presence of fecal matter but not necessarily the presence of pathogenic organisms.

Combined Sewer, carries municipal, industrial, and stormwater in a common pipe.

Denitrification, conversion of nitrate nitrogen to gaseous nitrogen, a normal atmospheric constituent.

Dissolved Solids, solids that pass through filter mats and require tertiary treatment for removal.

Effluent, liquid which emerges from a treatment process.

Eutrophication, the natural aging processes in a body of water.

Filtration, removal of suspended solids from waste water using adsorption and straining processes as the waste water flows through media such as sand or carbon.

Floc, a settleable group of solids formed in waste water treatment by chemical or biological processes.

Flocculation, process by which solids aggregation occurs.

Groundwater, water contained below the surface of the ground.

Heavy Metals, mineral elements such as mercury that are toxic in low concentrations to plant and animal life.

Incineration, combustion of sludge to reduce the volume and produce a sterile ash.

Industrial Waste Water, water containing pollution resulting from manufacturing processes.

Influent, the liquid that enters a treatment process.

Ion Exchange, process of removing dissolved inorganic mineral salts from waste water.

Lagoons, ponds in which sunlight, algae, and oxygen interact to restore water quality. The action may be speeded up by aeration with mechanical methods.

Land Treatment, generic term being applied to well-designed advanced waste water treatment systems where effluent is applied to soil on which crops are growing in order to use both the soil's filtration qualities and the crops' capabilities for taking up nutrients from the waste water as part of the treatment process.

MGD, flow rate in million gallons per day; 1 mgd = 1.5 cubic feet per second.

Municipal Waste Water, water containing pollution resulting from domestic wastes; typically feces and laundry wastes.

Nitrification, sequential conversion of ammonia nitrogen to nitrate nitrogen.

Nutrients, substances such as nitrogen and phosphorus utilized by plants in their life processes.

Organic, having molecular composition including carbon in combination with one or more elements such as hydrogen or oxygen.

Overland Flow, a type of land treatment in which the waste water flows slowly down a gentle slope over the soil surface through growing plants rather than infiltrating the earth. The effectiveness can be comparable to the infiltration method, and in certain climates under certain soil conditions, it has advantages over the infiltration process.

Oxidation, consuming or breaking down of organic wastes or chemicals in sewage by bacterial action or chemical oxidants.

Package Sewage Treatment Plant, a relatively small, prefabricated, self-contained primary-secondary treatment system. One might be used in a small subdivision where septic systems will not meet requirements for waste water discharge.

Pathogens, disease-producing organisms.

Physical-Chemical Treatment, renovation of raw waste water without the use of biological oxidation processes.

Pollution (water), results when something—animal, vegetable, or mineral—reaches water, making it more difficult or dangerous to use for drinking, recreation, agriculture, industry, or wild-life.

PPM, parts per million, 1 ppm = 1 part of the substance concentrated in one million parts of water (by weight).

Primary Treatment, removes large floating objects and settleable solids that are respectively screened and settled out of solution.

Refractory Organics, stubborn dissolved organic wastes such as detergents which resist biological oxidation.

Salts, minerals that water picks up as it passes through the air, over and under the ground, and through household and industrial uses.

Secondary Treatment, a basic step in sewage treatment following a primary stage of treatment, this one depending on the biological processes of bacteria breaking down organic matter. Includes activated sludge or the trickling filter process. A single lagoon provides both primary and secondary treatment.

Sedimentation, removal of settleable solids by gravity.

Settleable Solids, suspended solids that will separate by gravity under quiescent conditions.

Sludge, solid matter that settles in sedimentation tanks.

Solids, matter that remains as residue upon evaporation and drying at slightly above the boiling point of water.

Spray Irrigation-Infiltration, spreading waste water on soils where it is renovated as it passes through the soil and is acted upon by the plants.

Stormwater, rainwater containing pollution such as animal feces, chemicals, and refuse from streets and agricultural fertilizers and pesticides.

Suspended Solids, undissolved solids in a water sample that will not pass through a filter mat.

Tertiary Treatment, renovation of a biologically treated waste water by using chemical and mechanical systems. Though "tertiary" indicates a third stage, it could involve numerous steps in purification beyond a secondary stage. It is also called "advanced waste water treatment."

Toxic Organics, stubborn dissolved organic wastes such as pesticides and highly poisonous industrial chemicals like cyanide that resist biological oxidation.

Bibliography

American Chemical Society. (1969) "Cleaning Our Environment, the Chemical Basis for Action." Report by Subcommittee on Environmental Improvement, Committee on Chemistry and Public Affairs, American Chemical Society, Washington, D.C., 249 pages.

Anon. (1951) "Night Soil—U.S.A." *Industrial and Engineering Chemistry, 43:*15A.

Anon. (1971) "Special Report—Sludge Disposal: A Case of Limited Alternatives." *Deeds and Data,* Water Pollution Control Federation, Washington, D.C. (December, 1971).

Anon. (1970) "Wastewater Disposal Enhances an Area Ecology." *Industrial Water Engineering* (March, 1970), pp. 18–20.

Association for the Preservation of Cape Cod. (1972) *Waste Water Impacts: A General Survey of Cape Cod and a Detailed Analysis of Falmouth.* Report prepared by Bauer Engineering Inc., Chicago, in association with Anderson-Nichols and Co., Boston.

Bassett, James. (1972) "Ol' Man River: Cities Must Do Sumpin'." *Los Angeles Times* (September 18, 1972), p. 4.

Bernarde, Melvin A. (*circa* 1972) "Health Effects of Land Disposal of Treated Sewage Effluents: An Appraisal." Unpublished paper from author at Department of Community Medicine, Hahneman Medical College and Hospital, Philadelphia, 43 pages.

Bouwer, Herman. (1968) "Returning Wastes to the Land: A New Role for Agriculture." *Journal Soil and Water Conservation, 23:*164–69.

———; Rice, R. C.; Escarcega, E. D.; and Riggs, M. S. (1972) "Renovating Secondary Sewage by Ground Water Recharge with Infiltration Basins." Water Pollution Control Research Series, 16060DRV 03/72, 102 pages.

Butler, R. G.; Orlof, G. T.; and McGauhey, P. H. (1954) "Underground Movement of Bacterial and Chemical Pollutants." *Journal American Water Works Association, 46:*97–111.

Cold Regions Research and Engineering Laboratory. (1972) "Wastewater Management by Disposal on the Land." Cold Regions Research and Engineering Laboratory, Hanover, New Hampshire, 185 pages.

Conn, Richard Leslie. (1970) "Liquid Sludge as a Farm Fertilizer." *Compost Science* (May–June, 1970), pp. 24–25.

Cooley, Lyman. (1913) *The Diversion of the Great Lakes.* Chicago: Sanitary District of Chicago.

Crawford, A. B., and Frank, A. H. (1940) "Effect on Animal Health of Feeding Sewage." *Civil Engineering, 10:*495–96.

Crawford, Stuart C. (1958) "Spray Irrigation of Certain Sulphate Pulp Mill Wastes." *Sewage and Industrial Wastes, 30:* 1266–72.

Culp, Gordon L., and Culp, Russell L. (1971) *Advanced Wastewater Treatment.* New York: Van Nostrand and Reinhold.

Culp, R. L., and Moyer, H. E. (1969) "Wastewater Reclamation and Export at South Tahoe." *Civil Engineering, 39:*38.

C. W. Thornthwaite Associates. (1969) "An Evaluation of Cannery Waste Disposal by Overland Flow Spray Irrigation." *Publications in Climatology, 22:*1–73 (Elmer, New Jersey).

Day, A. D.; Stroehlein, J. L.; and Tucker, T. C. (1972) "Effects of Treatment Plant Effluent on Soil Properties." *Journal Water Pollution Control Federation, 44:*372–75.

Dennis, Joseph M. (1953) "Spray Irrigation of Food Processing Wastes." *Sewage and Industrial Wastes, 25:*591–95.

Department of Environmental Services, Government of the District of Columbia. (1971) *Water Pollution Control Plant of the District of Columbia.* Brochure ES-6 (11/71), 12 pages.

Eby, Harry J. (1966) "Evaluating Adaptability of Pasture Grasses to Hydroponic Culture and Their Ability to Act as Chemical Filters." *Management of Farm Animal Wastes* (Proceedings National Symposium, May 5, 6, and 7, 1966, pp. 117–20), Published by American Society of Agricultural Engineers, St. Joseph, Michigan.

———, and James, P. E. (1971) "Isotope Tracer Techniques

Used to Determine Flow Patterns in Tertiary Waste Water Treatment." From *Nuclear Techniques in Environmental Pollution*. Vilna: International Atomic Energy Agency.

Falk, Lloyd L. (1949) "Bacterial Contamination of Tomatoes Grown in Polluted Soil." *American Journal Public Health, 39:*1338–42.

Frink, C. R. (1971) "Plant Nutrients and Water Quality." *Agricultural Science Review, 9:*11–25.

Gilde, Louis C. (1971) "Engineering a Better Environment." *Mechanical Engineering* (March, 1971), pp. 40–44.

Gilde, L. C. (1969) "Waste Systems Tailored to Campbell Plants." *Food Engineering* (August, 1969).

Gray, J. Frank. (1956) "Agricultural Utilization of Sewage Effluent." 4-page unpublished paper obtained from author at Lubbock, Texas.

———. (1963) "Design and Management of an Irrigation System for Municipal Sewage Effluent." 5-page unpublished paper obtained from author at Lubbock, Texas.

———. (1959) "The Use of Reclaimed Waters in Agriculture." 4-page unpublished paper obtained from author at Lubbock, Texas.

Greenberg, Arnold E., and Gotaas, Harold B. (1952) "Reclamation of Sewage Water." *American Journal Public Health, 42:*401–10.

———, and Thomas, Jerome F. (1954) "Sewage Effluent Reclamation for Industrial and Agricultural Use." *Sewage and Industrial Wastes, 26:*761–70.

Harper, Horace J. (1931) "Sewage Sludge as a Fertilizer." *Sewage Works Journal, 3:*683–87.

Harvey, Clark. (1965) *Use of Sewage Effluent for Production of Agricultural Crops.* Texas Water Development Board, Report 9 (Box 12386, Austin, Texas 78711).

Heukelekian, H. (1957) "Utilization of Sewage for Crop Irrigation in Israel." *Sewage and Industrial Wastes, 29:*868–74.

Heuvelen, W. Van, and Svore, Jerome H. (1954) "Sewage Lagoons in North Dakota." *Sewage and Industrial Wastes, 26:*771–76.

Hill, David E. (1971) "The Purifying Power of Soil." *Frontiers of Plant Science,* published by the Connecticut Agricultural Experiment Station, New Haven (Fall, 1971), pp. 4–5.

Hunt, Henry J. (1954) "Supplemental Irrigation with Treated Sewage." *Sewage and Industrial Wastes, 26:*250–60.

Hutchins, Wells A. (1939) *Sewage Irrigation as Practiced in the Western States.* U.S. Dept. of Agriculture, Tech. Bulletin no. 675 (March, 1939).

Hyde, Charles Gilman. (1937) "The Beautification and Irrigation of Golden Gate Park with Activated Sludge Effluent." *Sewage Works Journal, 9:*929–41.

————. (1950) "Sewage Reclamation at Melbourne, Australia." *Sewage and Industrial Wastes, 22:*1013–15.

Imhoff, Karl. (1955) "The Final Step in Sewage Treatment." *Sewage and Industrial Wastes, 27:*332–35.

Jackson, Leon W. (1947) "Sewage Plant Sells Sludge and Effluent." *Engineering News-Record, 139:*122–24.

Johnson, James F. (1971) "Renovated Waste Water." Chicago: University of Chicago, Dept. of Geography, Research Paper 135.

Kardos, Louis T. (1970) "A New Prospect." *Environment, 12:* 10–27.

Latham, Baldwin. (1867) "A Lecture on the Sewage Difficulty." London: E. & F. N. Spon.

Lear, John. (1971) "Environment Repair: The U.S. Army Engineers' New Assignment." *Saturday Review* (May 1, 1971), pp. 47–53.

————. (1965) "The Crisis in Water—What Brought It On?" *Saturday Review* (October 23, 1965), pp. 24–80.

Little, Silas; Lull, Howard W.; and Irwin, Remson. (1959) "Changes in Woodland Vegetation and Soils After Spraying Large Amounts of Waste Water." *Forest Science, 5:*18–27.

Lubbock, Texas. (1971) *Report on Makeup Water for the Upper Canyon Lakes.* Published by Freese, Nichols, and Endress, Consulting Engineers, 811 Lamar Street, Fort Worth, Texas 76102.

Lunsford, J. V. (1957) "Effect of Cannery Waste Removal on Stream Conditions." *Sewage and Industrial Wastes, 29:* 428–31.

Martin, Benn. (1951) "Sewage Reclamation at Golden Gate Park." *Sewage and Industrial Wastes, 23:*319–20.

————, and McNulty, Dan L. "An Historical Review and Evaluation of the Water Reclamation Plant, Golden Gate Park, San

Francisco, Calif." Unpublished, undated paper from files of San Francisco Recreation and Park Dept.

McGauhey, P. H. (1965) "Folklore in Water Quality Standards." *Civil Engineering-ASCE* (June, 1965), pp. 70–71.

McKee, Frank J. (1957) "Dairy Waste Disposal By Spray Irrigation." *Sewage and Industrial Wastes, 29:*157–64.

McKee, Jack Edward and Wolf, Harold W. (1963) *Water Quality.* 2d ed. California State Water Resources Control Board (Publication 3-A), Sacramento, California, 548 pages.

McNulty, Dan L. (1960) "The Utilization of Digested Sludge as Fertilizer, Conservation and Disposal." Paper presented at Regional Conference, Northern California Sections, San Jose, California.

Melbourne and Metropolitan Board of Works Farm. (1971) *Board of Works Waste into Wealth.* 16-page report obtained from Board of Works, Melbourne, Australia.

Merz, Robert C. (1956) *Report on Continued Study of Waste Water Reclamation and Utilization.* Publication no. 15, State Water Pollution Control Board, Sacramento, California, 81 pages.

———. (1957) *Study of Waste Water Reclamation and Utilization.* Publication no. 18, State Water Pollution Control Board, Sacramento, California, 68 pages.

———. (1955) *A Survey of Direct Utilization of Waste Waters.* Publication no. 12, State Water Pollution Control Board, Sacramento, California, 75 pages.

———. (1959) "Waste Reclamation for Golf Course Irrigation." *Journal of the Sanitary Engineering Division — Proceedings A.S.C.E., 85,* SA6, pp. 79–85.

———; Merrell, John C.; and Stone, Ralph. (1957) "Investigation of Primary Lagoon Treatment at Mojave, California." *Sewage and Industrial Wastes, 29:*115–23.

Metropolitan Sanitary District of Greater Chicago. (*circa* 1970) "The Beneficial Utilization of Liquid Fertilizer on Land." Looseleaf booklet of material on sludge disposal by the Metropolitan Sanitary District of Greater Chicago, 100 East Erie St., Chicago, Illinois 60611.

Metzler, Dwight F.; Culp, Russell L.; Stoltenberg, Howard A.; Woodward, Richard L.; Walton, Graham; Chang, Shih Lu; Clarke, Norman A.; Palmer, Charles M.; and Middleton,

Francis M. (1958) "Emergency Use of Reclaimed Water for Potable Supply at Chanute, Kansas." *Journal American Water Works Association, 50:*1021–60.

Mitchell, George A. (1937) "Municipal Sewage Irrigation." *Engineering News-Record, 119:*63–66.

———. (1931) "Observations on Sewage Farming in Europe." *Engineering News-Record, 106:*66–69.

———. (1930) "Sewage Farm Displaces Filter Beds at Vineland, N.J." *Engineering News-Record, 104:*65.

———. (1924) "Studies of Outlets and Crops on Sewage Irrigated Areas." *Engineering News-Record, 92:*284–86.

Muskegon County, Michigan. (*circa* 1971) *The Muskegon County Wastewater Management System.* 16-page booklet published by Muskegon County Board of Commissioners, Metropolitan Planning Commission and Department of Public Works.

National Academy of Sciences. (1972) "Accumulation of Nitrate." National Academy of Sciences, 2101 Constitution Ave., N.W., Washington, D.C., 106 pages.

National Academy of Sciences-National Research Council. (1966) "Waste Management and Control." Publication 1400, National Academy of Sciences-National Research Council, Washington, D.C.

Niles, A. H. (1944) "Sewage Sludge as a Fertilizer." *Sewage Works Journal, 16:*720–28.

O'Connell, William J., Jr. and Gray, Harold Farnsworth. (1944) "Emergency Land Disposal of Sewage." *Sewage Works Journal, 16:*729–46.

Parizek, R. R.; Kardos, L. T.; Sopper, W. E.; Myers, E. A.; Davis, D. E.; Farrell, M. A.; and Nesbitt, J. B. (1967) *Waste Water Renovation and Conservation.* Penn State Studies, 23, Administrative Committee on Research, Penn State University, 71 pages.

Parkhurst, John D. (1970) "Wastewater Reuse–A Supplemental Supply." *Journal of the Sanitary Engineering Division— Proceedings A.S.C.E., 96,* SA3, pp. 7318–653.

Porcella, Donald B.; McGauhey, P. H.; and Dugan, Gordon L. (1972) "Response to Tertiary Effluent in Indian Creek Reservoir." *Journal Water Pollution Control Federation, 44:* 2148–61.

Rafter, George W. (1897) "Sewage Irrigation." *Water Supply and*

Irrigation Paper No. 3, U.S. Geological Survey, pp. 9–99.
————. (1899) "Sewage Irrigation, Part II." *Water Supply and Irrigation Paper No. 22*, U.S. Geological Survey, pp. 11–85.

Reynolds, Benjamin J. (1971) "An Ecological Blueprint for Today." Presented to National Environmental Health Association, 35th Annual Conference on Environmental Health, June 26–July 2, 1971, Portland, Oregon.

Roechling, Herman Alfred. (1892) "The Sewage Farms of Berlin." *Minutes of the Proceedings of the Institution of Civil Engineering, 109:*179–268.

Rolph, Bart S. (1962) "Golden Gate Park: A Salvaged Land—Irrigated With Reclaimed Water." Paper presented to the 14th Annual California and Pacific Southwest Recreation and Park Conference, Oakland, California, February 13, 1962.

Rudolfs, Willem. (1937) "Salvage from Sewage?" *Engineering News-Record, 119:*1055–57.

————; Falk, Lloyd L.; and Ragotzkie, Robert A. (1951) "Contamination of Vegetables Grown in Polluted Soil. I. Bacterial Contamination." *Sewage and Industrial Wastes, 23:*253–68.

————; ————; and ————. (1951) "Contamination of Vegetables Grown in Polluted Soil. II. Field and Laboratory Studies on *Endamoeba* Cysts." *Sewage and Industrial Wastes, 23:*478–85.

————; ————; and ————. (1951) "Contamination of Vegetables Grown in Polluted Soil. III. Field Studies on *Ascaris* Eggs." *Sewage and Industrial Wastes, 23:*656–60.

————; ————; and ————. (1951) "Contamination of Vegetables Grown in Polluted Soil. IV. Bacterial Decontamination." *Sewage and Industrial Wastes, 23:*739–51.

————; ————; and ————. (1951) "Contamination of Vegetables Grown in Polluted Soil. V. Helminthic Decontamination." *Sewage and Industrial Wastes, 23:*853–60.

————; ————; and ————. (1951) "Contamination of Vegetables Grown in Polluted Soil. VI. Application of Results." *Sewage and Industrial Wastes, 23:*992–1000.

————; ————; and ————. (1950) "Literature Review on the Occurrence and Survival of Enteric, Pathogenic, and Relative Organisms in Soil, Water, Sewage, and Sludges, and on Vegetation." *Sewage and Industrial Wastes, 22:*1261–81.

Sanborn, N. H. (1953) "Disposal of Food Processing Wastes by Spray Irrigation." *Sewage and Industrial Wastes, 25:* 1034–43.

Sawyer, Clair N. (1965) "Milestones in the Development of the Activated Sludge Process." *Journal Water Pollution Control Federation, 37:*151–62.

Schriner, Phillip J. (1942) "Disposal of Liquid Sludge at Kankakee, Illinois." *Sewage Works Journal, 14:*876–78.

Searles, S. S. and Kirby, C. F. (1972) "Waste into Wealth." *Water Spectrum* (Dept. of the Army, Corps of Engineers), *4:*15–21.

Sheaffer, John R. (1972) "Pollution Control: Wastewater Irrigation." *De Paul Law Review, 21:*987–1007.

———. (1970) "Reviving the Great Lakes." *Saturday Review* (November 7, 1970), pp. 62–65.

———. (1972) "Statement: The Ecological Revolution: Is There a Role for County Government?" Followed by responses from William D. Ruckelshaus, Ben Sosewitz, Gene E. Willeke, Lee Botts, David Zwick, and Charles V. Gibbs. *The American County, 37:*8–18, 131–33.

Skulte, Bernard P. (1953) "Agricultural Values of Sewage." *Sewage and Industrial Wastes, 25:*1297–1303.

———. (1956) "Irrigation with Sewage Effluents." *Sewage and Industrial Wastes, 28:*36–43.

Snyder, Charles W. (1951) "Effects of Sewage on Cattle and Garbage on Hogs." *Sewage and Industrial Wastes, 23:*1235–42.

Sopper, William E. (1968) "Effects of Sewage Effluent Irrigation on Tree Growth." *Pennsylvania Forests, 58:*23–26.

———. (1971) *Spray Irrigation of Sewage Effluent and Sludge.* Penn State University Institute for Research on Land and Water Resources, Reprint Series no. 27, 9 pages.

Spencer, B. R. (1944) "Sewage Disposal by Irrigation" (abstract). *Sewage Works Journal, 16:*655–56.

Steel, Ernest W., and Berg, E. S. M. (1954) "Effect of Sewage Irrigation upon Soils." *Sewage and Industrial Wastes, 26:* 1325–39.

Stone, Ralph. (1953) "Land Disposal of Sewage and Industrial Waste." *Sewage and Industrial Wastes, 25:*406–18.

Stoyer, Ray L. (1967) "The Development of Total Use Water Management at Santee, California." Proceedings International Conference on Water for Peace, Washington, D.C., May 23–31, 1967.

U.S. Congress. Senate. Committee on Public Works. Subcommittee on Air and Water Pollution. "Operation and Maintenance of Municipal Waste Treatment Plants." Committee Print, 91st Congress, First Session, Washington, D.C.: U.S. Govt. Printing Office, November, 1969.

U.S. Congress. Senate. Committee on Public Works. Subcommittee on Air and Water Pollution. Panel on Environmental Science and Technology. "Water Pollution Control Legislation. Waste Water Treatment Technology, Part 8." Hearings, 92nd Congress, First Session, Serial no. 92–H18, May 13–14, 1971. Washington, D.C.: U.S. Govt. Printing Office.

U.S. Congress. Senate. Committee on Public Works. "Regional Wastewater Management Systems for the Chicago Metropolitan Area." Committee Print, 92nd Congress, Second Session, Serial no. 92-24. Washington, D.C.: U.S. Govt. Printing Office, March, 1972.

U.S. Congress. Senate. Committee on Public Works. "The Cost of Clean Water," vols. I and II. 92nd Congress, First Session, Document no. 92-28. Washington, D.C.: U.S. Govt. Printing Office, 1971.

U.S. Congress. Senate. Select Committee on National Water Resources. "Water Resources Activity in the United States, Present and Prospective Means for Improved Reuse of Water." Committee Print no. 30, Washington, D.C.: U.S. Govt. Printing Office, 1960.

Van Vuuren, L. R. J.; Henzen, M. R.; Stander, G. J.; and Clayton, A. J. (1971) "The Full-Scale Reclamation of Purified Sewage Effluent for the Augmentation of the Domestic Supplies of the City of Windhoek." *Proceedings 5th International Water Pollution Research Conference* (July–August, 1970), pp. I-341–I-3219.

Viets, Frank G., Jr. and Hageman, Richard H. (1971) *Factors Affecting the Accumulation of Nitrate in Soil, Water, and Plants.* U.S. Dept. of Agriculture, Agricultural Handbook no. 413.

Warrington, Sam L. (1952) "Effects of Using Lagooned Sewage Effluent on Farmland." *Sewage and Industrial Wastes, 24:* 1243–47.

Whetstone, George A. (1965) "Re-Use of Effluent in the Future with an Annotated Bibliography." Texas Water Development Board, Report 8, Austin, Texas 78711.

Wilcox, L. V. (1948) "Agricultural Uses of Reclaimed Sewage Effluent." *Sewage Works Journal, 20:*24–35.

Wilson, H. (1945) "Some Risks of Transmission of Disease During the Treatment, Disposal, and Utilization of Sewage, Sewage Effluent and Sewage Sludge." *Sewage Works Journal, 17:*650–52, 1297–1300.

Wolman, Abel (1924) "Hygienic Aspects of Use of Sewage Sludge as Fertilizer." *Engineering News-Record, 92:*198–202.

Woodburn, John H. (1972) "My Town, My Creek, My Sewage." *The American Biology Teacher* (February), pp. 61–66.

Wright, Kenneth R., and Toren, Ralph. (1972) "Land Treatment of Sewage for Environmental Quality and Resource Cycling in Colorado." Paper presented to the Chemurgic Council—33rd Annual Conference, Washington, D.C., May 11–12, 1972.

Zemaitis, William L. (1971) "Water and Waste Water." From *Environmental Health,* New York: Academic Press.

Acknowledgments

In addition to the research indicated by the foregoing references, I had the great benefit of numerous interviews, most of them through visits with the interviewees, a few by telephone and correspondence. I wish to thank the following persons for their time and efforts in behalf of my firsthand research for this book:

Gil Aberg, Penn State University; J. B. Askew, San Diego, California; William J. Bauer, Chicago, Illinois; Tom Benge, Los Angeles, California; Herman Bouwer, Phoenix, Arizona; Patsy S. Clark, Grand Rapids, Michigan; Sandy Cooper, Boulder, Colorado; Russell Culp, South Lake Tahoe, California; Paul Dean, Beltsville, Maryland; Donald Deemer, Camden, New Jersey; Harry J. Eby, Beltsville, Maryland; Steve Fanning, Lubbock, Texas; Terrence Frost, Concord, New Hampshire; Walter O. Fuhrmann, Fredericksburg, Texas.

And Jerry Gardner, Chestertown, Maryland; Robert Gidez, Washington, D.C.; Louis C. Gilde, Camden, New Jersey; J. Frank Gray, Lubbock, Texas; Joe Haworth, Los Angeles, California; Fred Heilbrunn, New York City; Burd Hikes, Bartlett, Illinois; Edward W. Houser, Santee, California; James F. Johnson, Washington, D.C.; John Johnston, Austin, Texas; Louis T. Kardos, Penn State University; Glenn MacNary, North Falmouth, Massachusetts; Benn Martin, San Francisco Department of Public Works; Earl Meyer, Penn State University; Bud Nagelvoort, Washington, D.C.; Henry Owades, New York City.

And Donald Parmlee, Elmer, New Jersey; Richard R. Parizek, Penn State University; James D. Phillips, Colorado Springs, Colorado; Patrick J. Powers, Jr., San Francisco, California; W. Roper, Washington, D.C.; Gale Ruffin, Santee, California; John R. Sheaffer, University of Chicago; M. A. Simmonds, Brisbane, Australia; William E. Sopper, Penn State University; John J. Spring, Golden Gate Park, San Francisco, California; Pat Stoppe, Washington, D.C.; Ray Stoyer, Irvine, California; Frank G. Viets, Fort Collins, Colorado; Theodore C. Williams, Grand Rapids, Michigan; Gordon W. Willis, Lubbock, Texas; Kenneth R. Wright, Denver, Colorado; and Victor Younger, Riverside, California.

With my fullest appreciation,
Leonard A. Stevens

Index

283